Stress Free™ Work Process Solutions

i

Ron Mueller

Praise for
Stress Free™ Work Process Solutions

"The people matter! The people produce the products. The people fix the problems in the Supply Chain. The people can be developed and grown. Equipment and systems decay and must be maintained by people.
The people matter!"
REM

"Capable, well trained people working in a well-designed work process will win!"
GLM

Ron Mueller

Stress Free™ Work Process Solutions
For All Work Processes

By: *Ron Mueller*

Around the World Publishing LLC
4914 Cooper Road Suite 144
Cincinnati, Ohio 45242-9998

Copyright © 2021

ISBN 13: 978-1-68223-248-4
ISBN 10: 1-68223-248-4

Distributed by: Ingram
Cover Picture by: Suria Photo, Shutterstock, Inc.
Cover Design by: Ron Mueller

Ron Mueller

Technical Editor:
Gordon Miller P. E.

DEDICATION

To: All the hard working people, in all the countries of the world.

Table of Contents

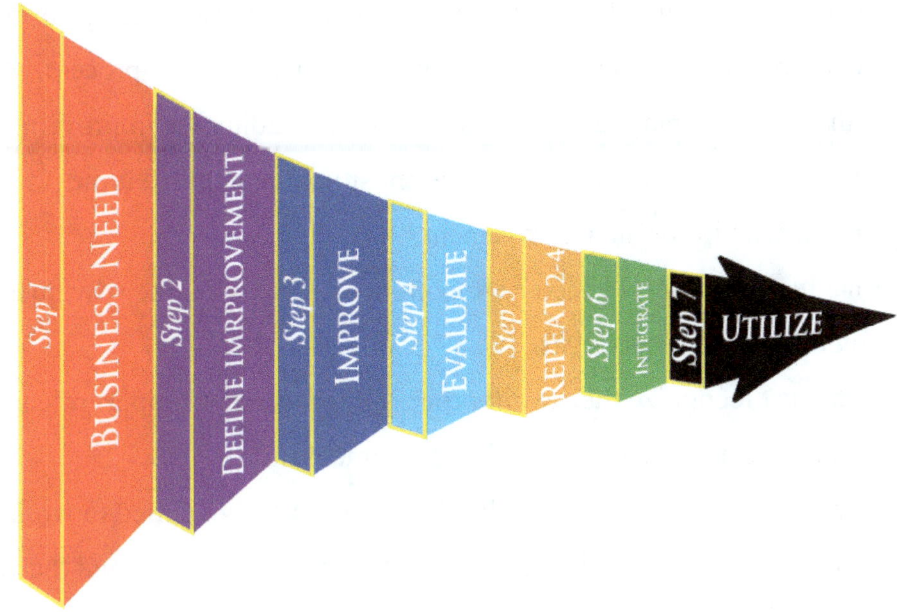

Introduction

This book is a guide to make work easier, less stressful, more effective and more efficient. The primary and most important person is the one doing the Value-Required (VR) and Value-Added (VA) work. Implementing this seven step approach will improve the work process experience and improve the resulting output. Waste will be reduced.

How to do the work, documented by those doing the work, is the way to stabilize and improve it. The documented, standardized work process becomes the basis for continuous improvement. It allows technology to be leveraged to create an even better environment.

Work accomplished easily will be done to the defined standard. When this easy-to-do way is visually displayed and documented, it is consistently followed. New people can quickly learn the standard way to execute the task at hand.

Clarifying and defining the specific actions that add value to the product and then eliminating the non-value-add (NVA) effort is a critical improvement activity.

The people directly involved in the transformation of raw materials or data into products and information are the key in improving the work process. Any approach that does not put them in the center and in leading the improvement will lack the deep understanding of the current situation and forfeit the subsequent ownership for the improvement.

External resources provide a very important role in asking the why questions and in providing new perspectives.

Leaders must recognize their work process improvement support role and provide encouragement, resources and improvement time.

The approach shared in this book has been successfully applied to work processes in every kind of work and applied in many countries around the world.

Stress Free™ Work Process Solutions and the companion Excel workbook provide a solid basis for successful and maintained work process solutions.

This approach is a bottom up approach. It guides the leader to get out on the floor and work with those doing the work. This on the floor leader is then in position to understand the work situation and participate in improving it. By definition the "floor" is where the work is being done. The "floor" is the place where raw materials are transformed into a product. It is the place maintenance work or manual assembly is performed. And it is in the office area where data is turned into information.

This critical relationship between the person who will experience the work day after day and the business area leader responsible for the delivery of production is crucial. It is the ingredient that creates ownership by those doing the work and a willingness to work to a documented, fixed process that is positioned to be continuously improved.

Ownership of the standardized work process should be by the person or team responsible for doing the work. The person that validates adherence to the standardized work process is the first level budgetary leader in that particular organization.

Well documented, visualized, standardized work enhances knowledge renewal and requalification. It is a very good tool in training new people and accelerates their ability to contribute to the rest of the team. The visualized step-by-step documentation provides a visual guide to the external part of the work. The work of the mind, a crucial part of the work, must be captured and made part of the documentation.

The culture is set and maintained by leaders. The best way to shape the culture is by doing. What is practiced and what is rewarded is what shapes the culture. Adherence to standardized work must be recognized and rewarded.

When standardized work is not the way people are doing the work, the leader must call for an improvement review. People always gravitate to doing work in the easiest way possible. Standardized work must continually strive to be the easiest way work gets done.

When hard work is required to meet regulation, quality or safety, all other ways of doing the work must be made harder or the required work easier.

The environment of an organization actively engaged in adhering to standardized work is invigorating. Leaders are on the floor supporting workers focused on making improvements. All of them are connected to the business and understand the work needs to be done only as fast as required by the customer. This is a mutually supportive environment where everyone is treated like an owner.

"Stress Free"

Stress Free!   ***Really?***

What makes this approach stress free?

1. Work processes become effective and efficient.
2. *Stress Free™ Work Process Solutions* closely guides the user in the creation and improvement of the work.
3. *Stress Free™ Work Process Solutions teaches* and leverages teamwork!

       Team Balance

       Team Value Add

The accompanying *Stress Free™ Work Process Solutions Workbook* is an excel workbook that provides tools, guides the user and creates a documented standardized work process.

The main focus of this book is to teach how to focus, how to clarify, how to fully understand work processes and how to set up a sustainable, friendly work area.

Stress Free™ Work Process Solutions – The Process

1	**Step 1: Business Need**
2	**Step 2: Define the Improvement.**
3	**Step 3: Improve**
4	**Step 4: Evaluate**
5	**Step 5: Repeat (Steps 2-4)**
6	**Step 6: Integrate**
7	**Step 7: Utilize**

This process is available in an Excel workbook. The workbook guides the user through one of three types of work process improvement. The workbook provides a way to organize the work process improvement effort. It is an aid to standardize the approach and provides a way to maintain the improvement.

The examples in this book are given in detail in the supporting Excel workbook.

Value of the Improvement

A key focus for improving work processes is the improvement value expressed in dollars per year. Work process improvement may be fun, but the solution must save money, it must provide a financial benefit. A key and first step is to understand the value of the improvement. This value should always be front and center for those working the improvement.

Defining improvement value is critical. The value of the improvement needs to be enough to warrant the improvement effort and cost expenditure.

The most appropriate way to state the value of an improvement is in **dollars per year**. Most business cycles and budgets are yearly. The decision to support a work process improvement effort must compete with other choices the resources could be working on.

Business Linkage

The dollar value per year, due to the work process improvement, **must be defined**. It is critically important to create this linkage. What specifically will be gained and how will this gain be measured when the work process is implemented.

General, high level improvement goals are important; - inventory reduction, speed to market, and net outside sales (NOS) gain. However, these are longer term and harder to realize.

Example of specific, clear, measures:

- 15 min package changeover down to 5 min.
- Work in Process (WIP) inventory reduction from fifteen times the customer "pull rate" to two times the customer pull rate.
- Manipulation line staffing reduction from 6 to 2

These are directly linked to the work process improvement.

System Layout

Immediately walking and doing a hand drawing of the system or documenting or obtaining the flowchart of the work being done, is an important activity that grounds the individual in the understanding of the physical system that they are trying to improve.

It is also important to understand the organizational structure that maintains the current situation. If the improvement change is to be maintained, the organization's leadership must understand how it will be sustained.

The system / work process lay out provides an understanding of;

- People placement and activities
- material placement and handling issues
- current behaviors and practices by individuals
- synchronization issues i.e. - long conveyance, bottlenecks, surge points
- Quality and Safety issues,
- Work balance issues

It is important for the improvement team to discuss what they have seen and understood from the, on-the-floor, walk or direct work observation and interviews.

The leaders of the area and the work process improvement team should step back and understand;

- The Customer,
- The Material Flow
- The Equipment
- The Human Connection - how the Work gets done.
- The Environment

Customer Connection

It is critical to get away from the thinking of only meeting the demand of the equipment.

Students of Japanese TPM theory will recall Takt. The Customer Cycle Time (Takt Time) is the measure that provides the direct customer need signal to the production system.

It is important to connect every production system improvement to the need of the customer. This connection defines the value of the improvement.

Generally, organizations tend to work faster and produce more than they need. This speed desire often leads to work process issues.

The customer cycle time should have no safety/buffer factors associated with it.

Available time:

Available Time is equal to Calendar or clock time **minus** planned activities that have blocks of time specifically calling for the production system to be down.

Available time should be used to help find the barriers to a customer focused production system.

Material Connection

The supply of the raw materials and the synchronized handling of the finished product are both critical in ensuring the flow of the production system.

Raw material quantity and placement is critical in optimizing the flow of the production system.

The logistics of bringing the materials to the line and returning material remnants for later use is a critical element the must be analyzed. Flow and time analysis is an important analysis to utilize.

Equipment Connection

Physical layout and material supply layout to the equipment are often related to specific problems or are due to previously experienced problems.

The problem may be the number of stops, or breakdowns or the amount of effort it takes to keep an area up and running.

There are many opportunities to better utilize the equipment at hand. For example:

- Create flow by the synchronization of filler, capper, and labeler and case packer by shortening the distances between the equipment to the optimum so the line can start and stop in a synchronous fashion and generate no scrap during a changeover.
- Minimize the raw material inventory in the production area; bottles, caps, KDF's by delivering at Takt or at a pitch.

The equipment layout needs to be drawn to scale. The time that a single product spends having value added and the time it spends traveling needs to be analyzed. The travel time then needs to be eliminated or shortened.

Change Over (C/O) time:

C/O relates to inventory reduction. Time spent changing from one SKU to another is non-value added. No sellable product is produced during the changeover. Inventory should only be held to maintain top level customer service. Inventory reduction is only possible when the system provides reliable flow. Reliable flow is required to support a customer focused production system.

The Human Connection

Once the initial production system drawing or work process flow chart is completed then the actions of the people involved in the production of the product should be studied and understood.

It must be noted that if the work process is being done by the human, the system drawing will be a flow chart showing the work of, writing, or analyzing, or evaluating etc. The "production system" is this flow chart that must be documented and verified.

Standardized Work;

Standardized Work is *the work done directly on the raw material, to produce a product to meet the customer need.* All other work being done is not considered standardized work. ***Standardized means synchronized to the customer.*** It forms the basis for making continuous improvement.

Many organizations produce to a production schedule. This schedule is the equivalent of the "customer". In many cases it is designed to cover the many problems experienced by the logistics associated with moving materials and finished product.

Within this concept people will work to the required standards as defined by safety and product quality requirements and that required by law.

For continuous improvement to be possible, the critical and repeatable work must be done to standard. The time for doing this work is normally synchronized with the cycle of the equipment that is being supported. It may also be a service that is synchronized with an accounting or other process cycle.

Stress Free^{TM} Work Process Solutions is the guide in solving problems of manual and office work. This applies to office work where computer software is a key contributor to getting work done. Computer software doesn't necessarily break, but it must be adjusted to be in synch with any changes to the work processes being used.

People and equipment do work;

- Both need to be maintained.
- Both break down.
- Both periodically need fixing.

People maintain, people fix, and people think and make improvements. People are the key.

Stress Free^{TM} Work Process Solutions *focuses* on the people working in the production system, on the equipment and maintaining material flow. It focuses on the work people do whether on the production floor, assembly area or in the office environment.

Following the ***Stress Free^{TM} Work Process Solutions*** guidance will lead to improvements in three to five work days.

You will always deliver a superior, easier way of doing the work.

If not contact: **Ron Mueller**

Work Processes and the Working World.

Every working person follows or implements a work process. Most of these work processes are documented in a very general way. Many do not have clear measurable accomplishment metrics. Many are done routinely and have no current documentation.

Work can be broken into two major kinds:

- Knowledge Work
  - o Frequently intangible result
  - o Mental process not visible
- Physical Work
  - o Tangible and measurable results
  - o Visible and public work behaviors

In both cases work is a process and it has a specific result.

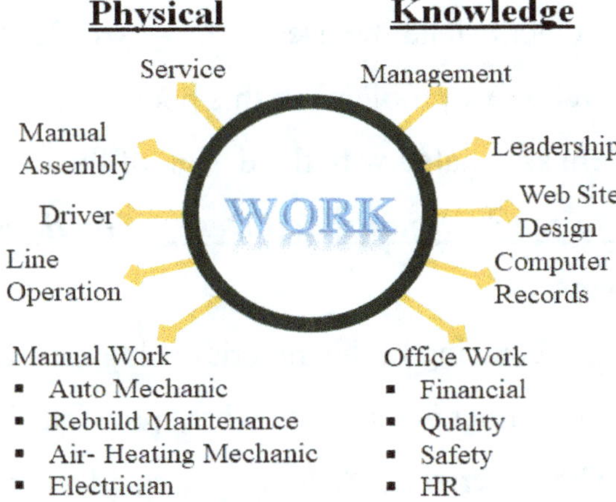

This is a simple example of the physical and the knowledge based work processes.

There are often overlaps or a mix between the physical and knowledge work.

These processes exist around the globe.

The country, the culture, the color of the people may vary but work processes exist everywhere. They are all quite similar.

<u>Work Improvement</u>

Work improvement implies that there are problems. Here are the six problem categories:

1. System Problems

System problems are problems experienced in the management, maintenance and assurance of the fundamental material to product transformation process. There are quality requirements, safety requirements, government and financial requirements. Adherence to these requirements is a fundamental expectation.

<u>Stress Free^{TM} Work Process Solutions</u> *will address many of these problems.*

2. Human Problems

The work related problems in this area are problems in how people work together. These problems lend themselves to very different problem solving tools. We will cover a few problems in this book.

The human problem associated with the design of the organization is covered in; ***<u>Stress Free^{TM} Organization Design Solutions</u>***.

3. Material Problems

These are problems occurring in the material being transformed. The material performance either is inadequate or it prevents the transformation from occurring. The material needs to be evaluated and action taken to ensure the material is within the specification required for easy and consistent transformation.

To do this please refer to: ***<u>Stress Free^{TM} Quality System Solutions</u>***.

4. Equipment Problems

These are problems occurring in the equipment. The equipment function is either reduced or stopped. Action must be taken to re-establish the transformation work the equipment is expected to provide.

To do this please refer to: ***<u>Stress Free^{TM} Manufacturing Solutions</u>***.

Work Improvement

5. Processing Problems

Processing problems are similar to equipment problems but often they are chemical reaction or flow and mixing oriented. Since visibility is often the problem, the analysis tools for these problems aid the investigator by providing visibility of the problem via secondary measures, mathematical models or physical models.

To do this please refer to: ***Stress Free™ Manufacturing Solutions.***

6. Information Problems

Information flow, timeliness, accuracy and understandability are critical in holding complex transformation processes together. It is often the single most significant problem in complex transformation processes. There are many ways information can flow;

- Visually – red light to stop cars, signs, symbols, a smile or frown
- Electronically – internet, e-mail, phone.
- Paper – reports, certificates, orders, receipts, lists
- Audibly – voice, horns, music

Usually information flow is a mix of the media described. Problems often arise due to incomplete or misinterpreted information. Problem solving relies heavily on the clarity of information. All the books highlighted rely on the clarity and correctness of information.

Every one of these problem types have been experienced and solved in every part of the world. They are universal in nature.

The state of work processes

Varieties

- Undocumented
- Documented but not followed
- Work process with no defined metrics
- Inter-related, dependent, uncoordinated

Documenting and defining the evaluation measure for each work process is critical. They work for pride first but people will do what they believe is necessary to satisfy their bosses and the "system".

Effectiveness

One VP at a global company developed The Million Dollar Loss report to be generated weekly. The Million Dollar is an example of a very powerful but ineffective, poorly aimed tool. Great name but when examined, it made no sense. It was however, expected to be on the VP's desk every Monday morning. He used it to brow beat various production plant managers about their poor performance.

Because of their dollar expenditure, large plants, even when they were performing well remained on the loss report list. Small plants though they might perform poorly seldom made it into the Million Dollar report.

I stopped compiling the report to see what would happen. Three months later, the VP called wondering about the Million Dollar report.

Once he understood the situation and got over the fact that I had unilaterally stopped compiling the report, he began using better metrics to evaluate the various sites. These more focused measures put the right production plant managers in front of his desk for additional guidance.

Documentation

Are your work processes documented?

Is the documented work process the actual one that is followed?

Are the performance measures for the work process clear?

Are the inter-relationships with other work process clear?

These questions will guide the improvement process.

A Global Look

Let's take a global trip and examine work processes with the mindset that this improvement work is applicable globally. It has been done in more than sixty countries.

But let's have some fun and learn a bit about each country.

Thailand

Serious Facts
Capital: Bangkok
Population: 66,000,000
Ranked: 50th in size of population
Population Growth – 3 million in 6 years

Fun Facts
When traffic is bad in Bangkok, the sky train quickly fills with riders eager

to by-pass the hours of delay in taxis, cars or buses.

Best Beer: Singha

<u>**Thai Work Process Problems**</u> **(3)**

1. <u>**Work on automated line**</u> – More value add work

The request was for more productivity from an existing production line. Often productivity means cutting people. This is counter to my thinking. I will seldom work on cutting people unless the competitive viability of the business is on the line. This to me means we will go out of business if more productivity is not achieved. Getting more profit margins at the expense of the worker is counter to my thinking. Fire about six vice presidents and you will get a terrific productivity boost throughout the organization.

A small team observed the line and provided solutions yielding significant improvement in throughput with the same staffing.

How it was solved?

The Pareto Interview Process (PIP) in Step 1 of the Seven Steps of *Stress Free^{TM} Work Process Solutions* provided new insight to the work process problems.

2. <u>**Standard Work on Sashay line**</u> – 9 people to 2 people solution

Nine operators were stationed to operate nine sashay shampoo filling units. The nine units were packed out as a single line. On the floor observation and detailed documentation of the value add work indicated that less than three people could accomplish the tasks.

A small improvement team was assigned to evaluate and to devise a more efficient way to operate this line. In four days the team devised and implemented a way to operate the line with less than two people.

In step 3 Improve, the site manager and the area manager demonstrated that two people could indeed keep the line running in a "stress free" way. This demonstration by the leaders made a significant impression on those operators normally doing the work.

How it was solved?

By using the Seven Steps of *Stress Free^{TM} Work Process Solutions.*

3. Standard Work in Customization - 47 to 16 solution

Customization of production is on a dramatic increase in the consumer products business. Consumers desire a more convenient mix and presentation of products. This desire causes the packaging of the product to take on many variations. Each variation may require special manipulation. This manipulation often requires manual intervention of already automatically packaged material.

The site was preparing to spend several million dollars to expand the customization area and double the staffing. The request for the improvement work session was to evaluate how many people and how much space would be required for this expansion of customization.

The improvement team evaluated the current customization process. In Step 2 the team identified the value added work. They modified the work process. In four days it demonstrated a reduction from forty-seven people down to sixteen people to accomplish the same production throughput.

The expansion and additional hiring was cancelled. The reduced staffing requirement meant the projected increase in customization could be handled with current staffing.

<u>Two additional savings were identified;</u>

1. a reduction from sixteen to twelve people by improved raw material presentation, packaging and
2. Transport loss elimination by moving customization to the end of the production line.

How it was solved?

By using Seven Steps of *Stress Free^{TM} Work Process Solutions.*

Romania

Serious Facts

Capital: Bucharest

Population 22,200,000

Ranked 80 in land size

Population Growth – (-100 thousand) in 6 years

Fun Facts

Bucharest is a great city to walk in. It has great restaurants and good food.

Best Beer: Timisoreana

Romanian Work Process Problem (1)

1. Bottle Blowing Extruder Centering – Immediate action

This problem was causing the site to generate a large amount of scrap at the start of the bottle blowing process or when bottle sizes were changed. The plastic extrusion called the "sock" had to have an even wall thickness all the way around.

The set up consisted of centering the central guide to ensure an even thickness of melt would get extruded from the nozzle. Multiple adjustments were needed to get the thickness even all around the bottle sock.

"It is an art to get the sock to come out right the first time," was the comment of the line leader as we watched the setup procedure.

The improvement team watched three setups. The best set up took a minimum of three adjustments. One took eight adjustments. Every adjustment resulted in at least three lost bottles. Every adjustment and evaluation took at least fifteen minutes.

There was plenty room for improvement.

The team applied mistake proofing discussed in Step 3: Improve and developed a special tool. Then, working with a key operator, the improvement team demonstrated a one step, one adjustment, and mistake proof procedure in the afternoon of the following day.

The result was zero lost bottles.

How it was solved?

By using the Seven Steps of *Stress Free^{TM} Work Process Solutions.*

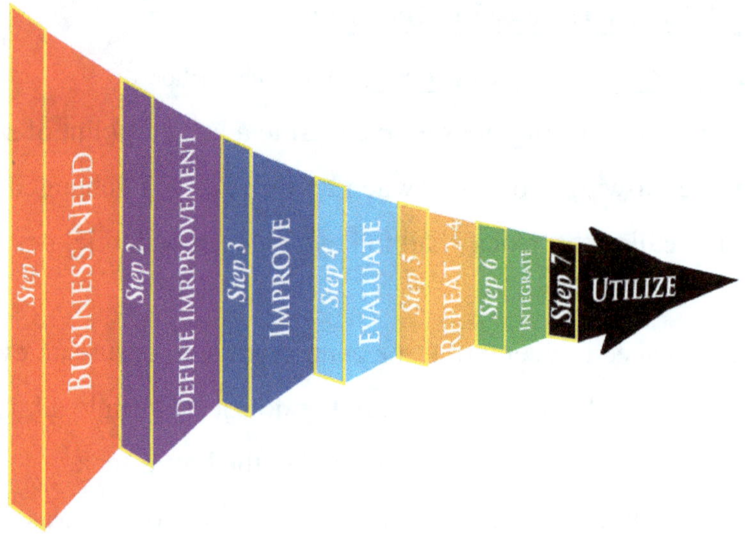

<u>Mexico</u>

Serious Facts

Capital: Mexico

Population: 110,000,000

Ranked: 14 in land size

Population Growth: 6 million in 6 years

Fun Facts

The country of the honorable cab driver thief. He charged three times the rate but provided three receipts!

Best Beer: Corona with a lime

Mexico Work Process Problems (2)

1. Blocked Payments

The purchasing office was experiencing blocked payment issues with several top material suppliers that threatened to impact the ability to produce the final product in a timely manner at the quantities desired.

The process seemed to be the problem of the entire organization and all suppliers. An improvement team, utilizing the tools of *Stress Free™ Work Process Solutions* focused the effort and resolved the issue in a matter of a few days.

The problem was consistently occurring with three suppliers and with one person in the purchasing office. The fact that the documented procedure was outdated and not followed was also a key factor.

The team, along with the one person experiencing the most problems, flow charted and tested an effective payment process. The flow chart was refined over the next week and final documentation entered in the organizations standards file.

The following quarter there were no blocked payments.

How it was solved?

By using the Seven Steps of *Stress Free™ Work Process Solutions*.

2. Core Waste elimination – Take it to the core

This was an embarrassingly simple, immediate $500,000 per year to the bottom line on the same day as observation. It is included here to demonstrate that work process and equipment problem solving relies on understanding the situation.

The improvement team went to the floor to observe the operation of the tissue combining stand. This transformation combined two sheets of tissue to create a double layer. Imagine a roll of toilet paper that proportionally is ten feet wide.

The team observed the large paper roll being unwound. As the paper came closer to the core the operator watched it get to the metal retaining ring holding the large roll on the machine. Then just before the unwinding paper got to the retaining ring the operator initiated the start of the new roll.

There was almost a foot of paper still on the core. When asked about this, the operator's response was sensible. The interaction of the paper with the retainer ring often caused the web to tear. When this happened the broken web went through the entire unwind stand. Then the web needed to be totally rethreaded. This rethreading often took up to thirty minutes. By stopping it early and making a controlled cut, the end of the new roll and the tail of the old web could be tied together. The system then would rethread automatically in less than a minute.

This was why the site continued to allow almost a foot of paper to be left on the core and "recycled" back to papermaking.

"Why is the retaining ring so large?" I asked the improvement team.

"So the core can be held solidly by the machine," was the reply.

"What if the retaining ring was the exact size of the core outer diameter?" was my question.

A spare set of retaining rings were immediately taken to the machine shop and modified. On the next paper roll change the new ring was put in place. The operator was asked to take the paper down to the edge of the metal retaining ring. No change in the work process.

There was no problem and the core and less than a half inch of paper was removed from the machine. The new retaining rings were tested for the rest of the day. By the next morning the retaining rings on all combining machines had been modified to the new design.

In this case there was no need to retrain any of the operators.

How it was solved.
By using the Seven Steps of *Stress Free*[TM] *Work Process Solutions*.

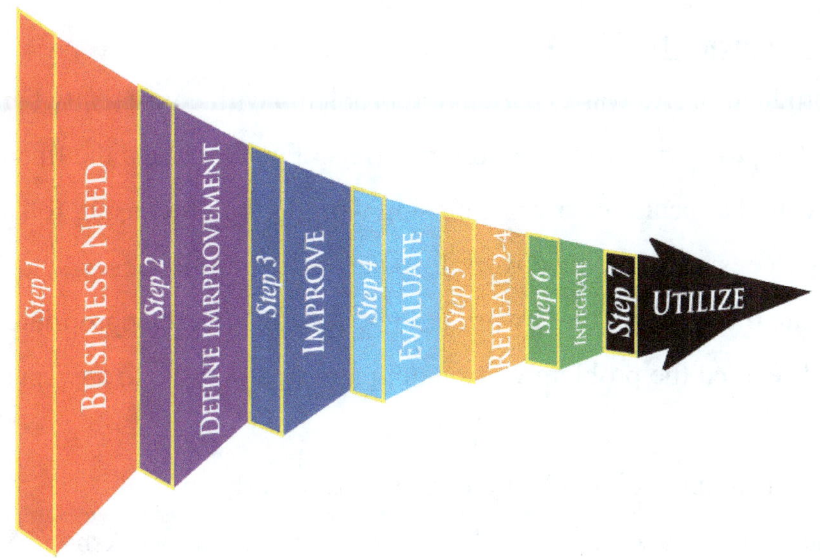

<u>United States</u>

Serious Facts

    Capital: Washington, DC

        Population: 326,474,013

        Ranked: 3 in land size

        Population Growth: 23 million in 6 years

Fun Facts

    The "why should we do it" country. This has been the hardest country for me to get folks to do it better.

    Yet by far the best country in the world to live.

    It fulfills the saying "It is hard to get the good to be better."

Best Beer: Bud Light.

US Problems (2)

1. <u>Late payments by Finance</u>

The production site was experiencing problems with consistently being on time with the payment for production materials. Each morning the finance manager would allocate the invoice closing work to the purchasing folks. She was trying to ensure a balanced work load across her employees.

Yet quite often the required work did not get done in the appropriate amount of time. It seemed the problem was random and was not due to any specific individual.

A request for an additional employee was turned down.

Her next step was to hold a work process improvement workshop. She had experienced a *Stress Free™ Manufacturing Solutions* workshop and wondered if there was something similar for the office area.

She sponsored a *Stress Free™ Work Process Solutions* workshop. Her three improvement teams investigated the Materials Payment process, closing the books each month and closing the books at the end of the year.

By the end of the week long workshop all three processes were improved. The time to close the books each month went from more than seven days to two. The time to close the books at the end of the year went from two weeks to less than a week. And the problem of managing the payment for materials became one of the novel solutions. It was very visual.

The Payment process was flow charted so that everyone was doing the work the same. The finance manager came in each morning and allocated the work as she had always done.

The difference was when any of her staff ran into a problem, they would raise a red flag, and yes a real red flag like the one you would have on a bicycle. This meant they had run into an issue that was taking more time than allocated for the payment process. The manager would then re-allocated the work and help in handling the payment processing.

The visual nature of the problem also allowed the other staff personnel to volunteer help when they were ahead.

Late payments went to zero, the office area camaraderie increased and the suppliers became more satisfied. This is a true, win, win, and win result.

How it was solved?

By using the Seven Steps of *Stress Free™ Work Process Solutions.*

US Problem

2.   Parent Roll Centering - put it in the center of the road

Paper hand towels and toilet tissue have a very complicated production process. Additionally, the equipment is very large. The parent roll centering problem was a chronic problem in the converting process. This process takes a huge roll of paper called a parent roll and transforms it into the familiar paper towel rolls and the bathroom tissue rolls. The problem had been experienced from the first day the converters of this kind were installed.

The problem as stated was that papermaking was making bad rolls and passing them on to converting. The converting folks were convinced the converter could not convert the loosely wound parent rolls.

The improvement team went out to the floor to observe the problem. A "bad" roll was brought to the line and mounted as per the standard loading procedure. The converter could not convert the roll because the parent roll telescoped as it ran.

The line leader was asked to mount another bad roll.

"It will do the same thing." was the line leader's response.

"I believe you but the improvement team needs to see it again," was my reply.

The improvement team was instructed to read the standard loading procedure as the process was being executed.

Bad roll number two was loaded as per the standard and it telescoped and the web broke.

"Does the team understand the problem," was my next question?

In the work process improvement team meeting room, the team put together their cause and effect diagram and came up with only a few causes.

One process cause was so obvious they immediately went out to run one more trial for understanding. They stopped by the machinist shop to pick up a magnet held red laser.

"I know, you want to load another bad roll," was the line leader's comment as the team came back out.

This time with the improvement team guided the loading of the bad roll; the converter converted the entire roll. They had the operator position the roll so that the laser mounted on the frame illuminated the end of the parent roll when it was in the optimum position.
The roll was positioned so that the strategically aimed laser turned edge of the parent roll a bright red. The operator was to keep the red on the end of the roll maximized.

 The roll ran flawlessly.

"You guys were lucky," was the comment of the line leader.

The team was so confident of their solution they did two more "bad" rolls.

The work process of loading and centering the roll was modified. Every operating team was trained on the new procedure. The line performance increase significantly.

The line leader became the best advocate for the solution.

The solution was achieved on day 1 of a three-day problem-solving session. It is highlighted here because it was a combination of work process and equipment problem solving. This is often the case.

How it was solved?

By using the Seven Steps of *Stress Free*^{TM} *Work Process Solutions.*

Chapter 1 Business Need

Step 1: Business Need

Step 1.1 Understand the Work Area
- Material Transformation Analysis (MTA)
- System Diagram
- Top Level Flow Diagram
- Physical Layout Diagram

Step 1.2 Problem Aiming and Stratification

Step 1.3 Understand the Customer

Step 1.4 Area Self-Assessment

Step 1.5 Observe Selected Priority Area

Step 1.1 Understand the Work Area

Material Transformation Analysis (MTA)

Material Transformation Analysis is an adaption from industrial engineering that originally focused on people and their actions. Only two elements do work; equipment and people. Equipment ages and declines in its work ability. People when nurtured grow and become more valuable. They think, they do, they take care of the equipment. Enough cannot be said about developing the individual to the point where the individual becomes the catalyst for continuous improvement.

The benefit of MTA is how quickly it is done. A complex system quickly becomes understandable. Entire production lines fit on an 8.5 by 11 piece of paper. Training is improved. Problems are quickly identified. Improvement priorities are visible to every associate on the line.

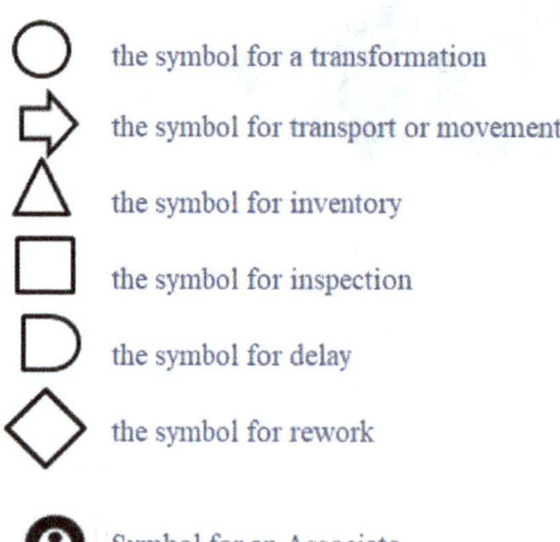

Standard Symbols used

These symbols are standard for the Industrial Engineering area. Transformation is used in the broad sense for data and raw materials. Transport is used in the same broad sense. Data is transported electronically, where materials are transported physically. Inspection can be done for the material transformation and as well for a data transformation into information.

Example 1: Hamburger Making Line

Hamburgers were prepared, frozen and boxed for shipment. The production line had a variety of problems. Each Associate team member had set of problems they felt were most important.

Each line associate was asked to create a list of their problems. This list got discussed and prioritized. Improvement occurred on a daily basis.

Material Transformation Analysis – Hamburger Making Line

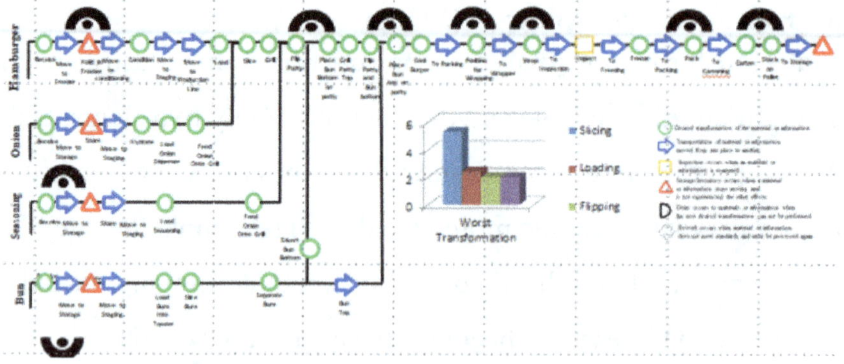

At least one major improvement was accomplished at every station. The production line up time and throughput went up approximately 20%.

This was accomplished in a five day workshop.

Example 2: Custom Hand Pack line

This custom hand pack line was maxed out in its capacity. It was not keeping up with demand. Additionally there was a demand of the customization of additional product.

In this case, the forty or so associates were often a different group mix as they went day to day and week to week. The manaagement team were permanent employees. The initial transformation analysis was done by the management team.

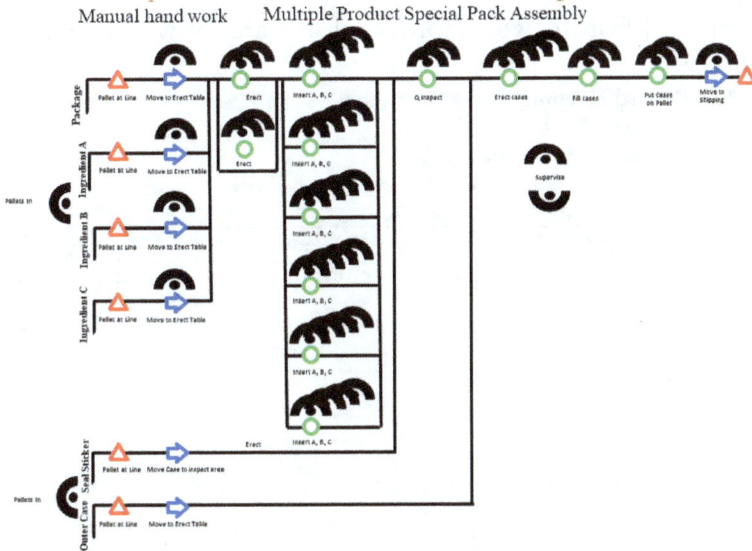

The key improvements were to make permanent shift teams and to train members at each station. Each member qualied at three stations.

This was accomplished in a five day workshop.

Example 3: Accounting / Finance Area.

1. Primary Processes

Notice that the transformation analysis is now listed vertically with fewer symbols. The transformation is highlighed but the action or steps in between are simplified. This could use all the symbols used in the previous two examples but this simpler approach proves sufficient for many office work processes.

The value add flow analysis is also very effective in the office environment and is often used to complement the MTA.

The steps 2-4 For the work porocess listed are shown in detail in the Excel Workbook.

- Record Keeping
- Auditing
- Additional Work Processes

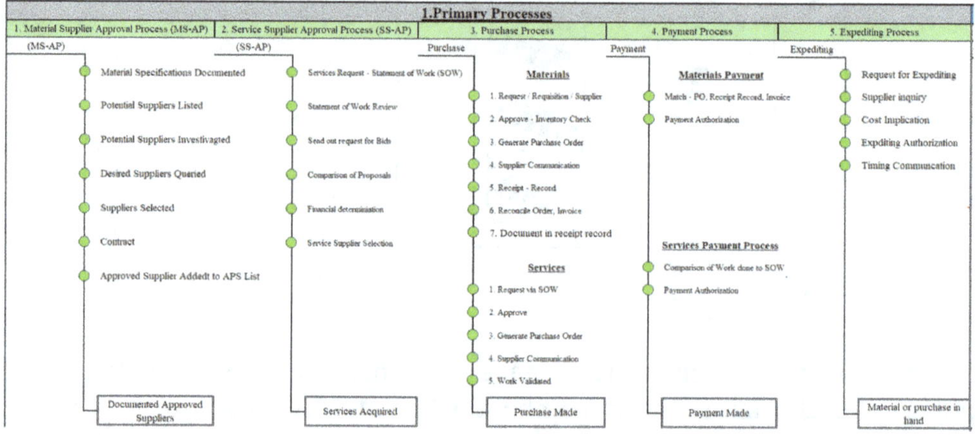

This Transformation analysis provided the information to standardize common repetitive work. This in turn made it easy for the finance personnel to support each other and make work flow smoothly. Blocked payments went from being the number on problem to zero.

Improvement was accomplished in a five day workshop

Use of the System Diagram

The system diagram can be very simple as shown here or it can get much more detailed and complex. Start simple and add more detail if it is needed.

This is the hamburger making line shown as a simple flow diagram.

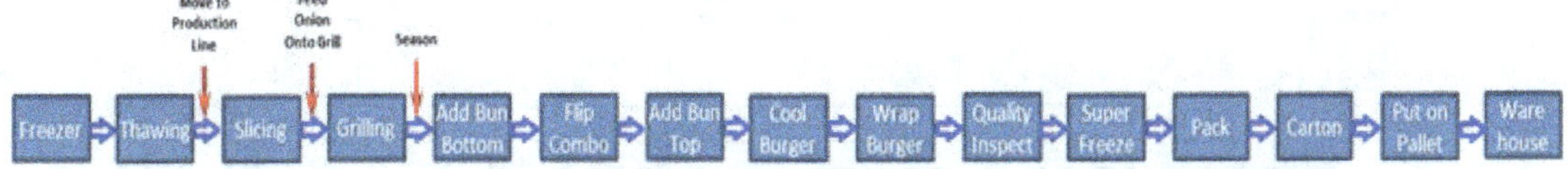

Top Level Flow Diagram for the financial area.

This is a financial area flow diagram of the work done in that office.

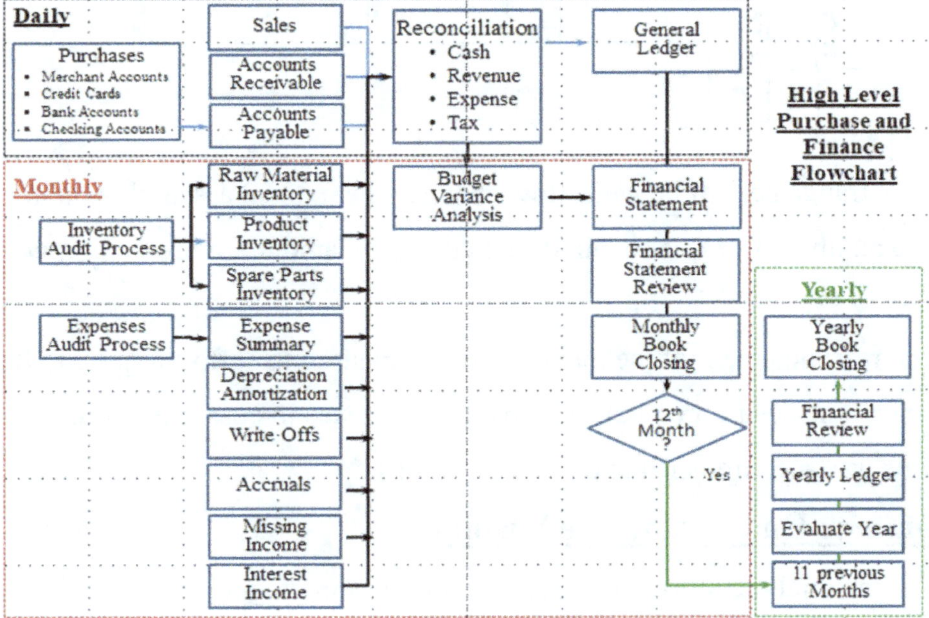

Physical Layout Diagram

The physical layout out diagram provides another important way to understand the work area.

This is the layout of the customization area.

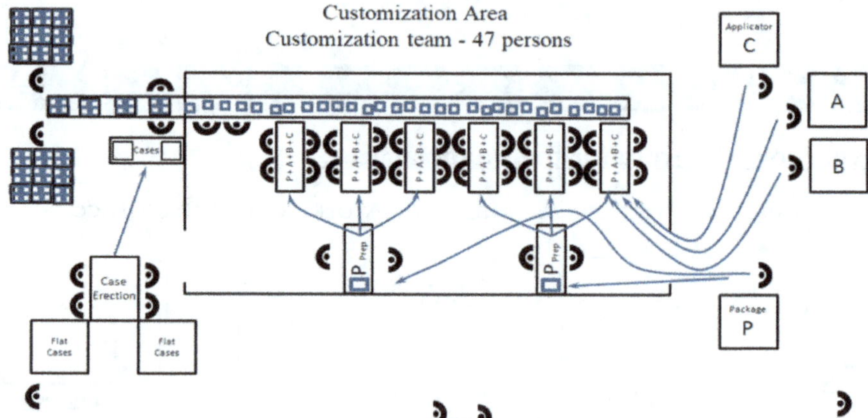

A rearrangement of he work flow changed the manual assembly in a manner that the number of people required to do the same amount of production went from 47 to 14.

The rearrangement of the manual work decreased the effort required. It also eliminated the need for a planned expansion that was to cost two million dollars.

This was accomplished in a five day workshop.

Summary: Understand the business

The material transformation analysis, the system diagram, the top level flow diagram and the physical layout diagram each provide a level of understanding that provides all improvement participants a deep level of understanding that gets used to stratify and aim the improvement effort.

Step 1.2 Problem Aiming and Stratification

Production Systems and Work Processes experience many problems. Often various leaders and work associates have differing opinions about which problems should be addressed and in what order.

Step 1.1 provides a common understanding to the improvement team. However, the deeper understanding and team discussion normally leads to the identification of more improvement items than is possible.

There is a very quick and accurate way to utilize the knowledge of the individuals doing the work to get to the actionable elements of the problem and the "Hard Work".

Pareto Interview Process (PIP)

Step 1: On a chart pad draw the first bar of a Pareto.
Step 2: Ask: What is the second worst problem?
Step 3: Ask: What is next worst problem?
Step 4: Repeat the process for each of the three Pareto bars in step three.
Step 5: Select the problem at the lowest level that the people in the room choose.

Pareto Interview Process (PIP)

The **Pareto Interview Process (PIP)** works very well for Identifying the "Hard Work". The hard work is known by those doing it. However, the main problem is zeroing in on the actionable elements of the problem. Every participant will have a slightly different view and will normally mix in fixes for the problems they see.

The **Pareto Interview Process (PIP)** for a specific problem is an interview technique that relies on the knowledge of the people in the room. Five knowledgeable people will correctly aim the improvement focus more than 85% of the time.

The facilitator has the simple task of asking two questions;

1. What is the worst problem you experience?
2. What is the next worst one and how does it compare to the worst one that ranks higher?

This initiates the aiming process.

Step 1: On a chart pad draw the first bar of a Pareto.

- By definition this problem is the worst and gets a 100% value.

Step 2: Ask: What is the second worst problem?

- Ask: Relative to the first how bad is the second problem?

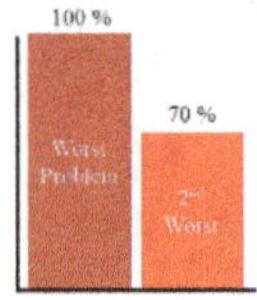

- Draw this out on the chart pad.

Step 3: Ask: What is next worst problem?

- Often the third problem is a significantly lower issue. There may be problems beyond the fourth or the fifth bar but normally they are of even lower value.

Step 4: Repeat the process for each of the three Pareto bars in step three.

- Go down another two levels if possible.

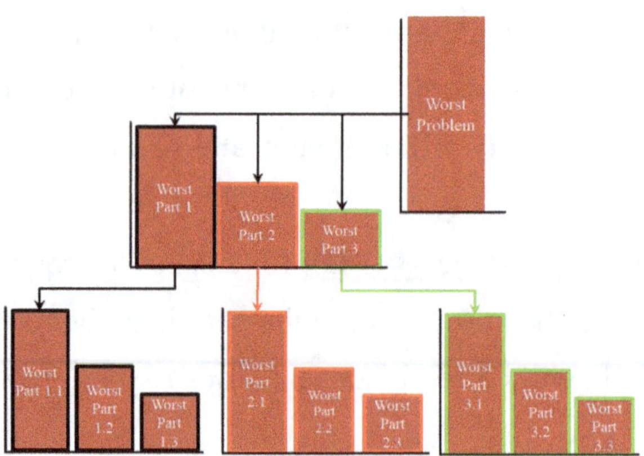

Step 5: Select the problem at the lowest level

Here is the resulting Excel worksheet showing the stratification

A verbal description of the situation is included. This clarifies the problem, its value and how the resulting solution will be evaluated.

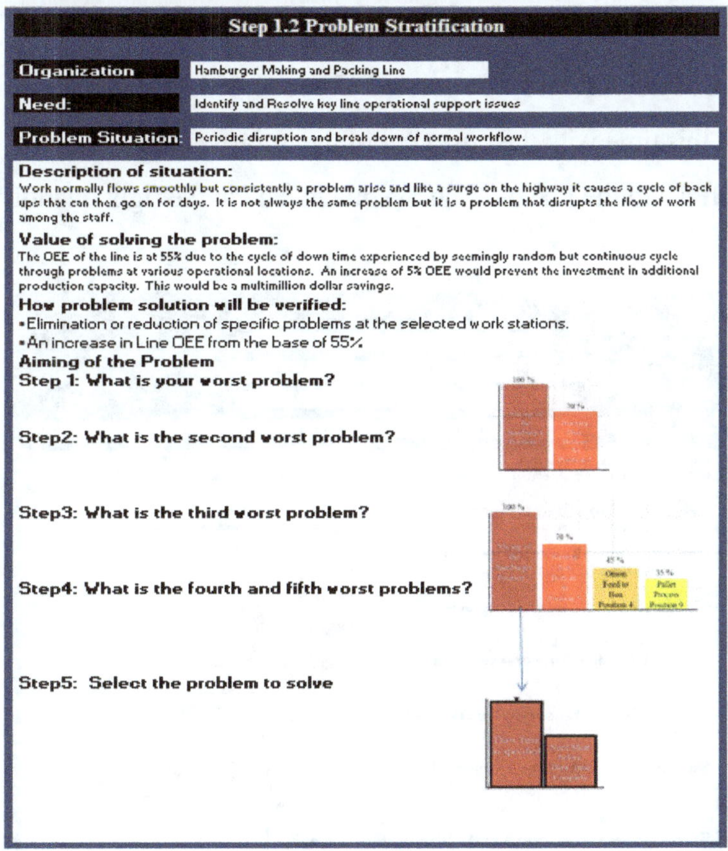

Step 1.3 Understand the Customer

The customer is the next in line to receive transformation output. The customer may be internal or may be external. The customer in this case is the line operating team. The supplier is the improvement team.

The same **Pareto Interview Process (PIP)** used with the organization is once again used but this time with the customer of the improvement. The problem of getting the material to the line on time, on quality and at the right quantity is the defined need.

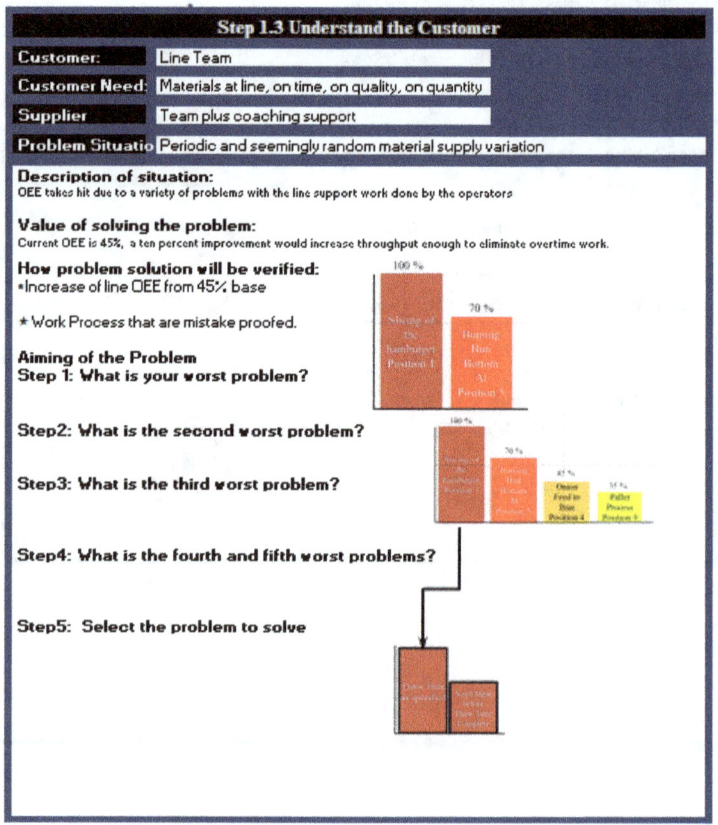

Step 1.4 Area Self-Assessment

People with drive overcome hard work. They however look to lessen, or eliminate the hard work. They look to eliminate the grind. The part of the work that makes it hard. How you handle the grind is often what separates the winners from the quitters. If hard work is to be worthwhile it must meet certain standards and measures that matter to those performing the work.

Creating an improvement plan gives everyone involved an opportunity to contribute, or object to aspects of the change. Hard work is a day by day issue. The improvement needs to address the work and how it contributes to the need of the business.

Why does this work need to be done?

Is it value added work?

Is the current struggle worth it?

Identifying the improvement and making it happen quickly is important. Doing the "right thing" must be easier than doing the "less than the right thing". Everyone with an oar in the water must pull together.

When one thinks of one's self it is of the person that we were when forty or thirty or twenty. When you look in the mirror you are often surprised at the person looking back.

The work of the organization has a similar analog. Everyone sees the work in a different way and will usually be surprised at the variations by which the work gets done.

And finally when the work is monitored and freshly documented the surprise is that it is different than almost everyone expected.

The *Stress FreeTM Work Assessment* process facilitates the organization to get a good qualitative understanding of the current work situation.

Thirty-five questions covering the customer, the work processes, the material flow, the people, the support systems, the environment and leadership provide a comprehensive qualitative assessment of the current work situation.

Work Area Self-Assessment Questions

Work Area Assessment Questions		No	Not Sure	OK	Mostly	Good	Best in Class
Customer							
1	Customer Satisfaction is the primary focus of the work				a		
2	Customer quality expectations are exceeded			a			
3	Customer input is sought and utilized					a	
4	Customer Satisfaction is rated after every interaction				a		
5	Frequent customers are utilized to improve the business				a		
Work Processes							
6	Work Processes are easy to learn and do.			a			
7	Work Processes are visualized and mistake proofed		a				
8	Each major work process has been optimized			a			
9	Work design ensure easy to maintain practices			a			
10	Work processes get periodical review and improvement focus			a			
Material							
11	Material Quality Characteristics have units and tolerances defined.			a			
12	The supplier is certified and the material meets the required specifications					a	
13	Incoming materials are verified to correctness, quality and quantity.					a	
14	Damaged incoming materials are not accepted				a		
15	In process material handling issues are rare and easy to recover from.			a			
Human							
16	Expert problem prevention skills are the norm.			a			
17	Advanced condition management skills are the norm.		a				
18	Adherence to Standards and following Standard Operating Procedures is the norm.				a		
19	Mistake proofing has eliminated mistakes.	a					
20	Our people are highly motivated, continuous improvement leaders				a		
Systems							
21	Systems exist and are well documented.			a			
22	Systems ensure Cpk>1.33 and no defects				a		
23	Systems ensure required production rate is maintained.				a		
24	Systems are explained and trained using the lastest techniques			a			
25	Systems are easy to learn requiring less than five days to get qualified.		a				
Environment							
26	Support Organizations accept our solutions and integrate it into their design.				a		
27	Purchasing adjusts its buying based on the problem solution.					a	
28	Material specifications are adjusted based on specific problem solutions.					a	
29	Operational conditions are changed as required to maintain improvements				a		
30	Logitics are adjusted to support specific problem resolution.					a	
Leadership							
31	Leaders review the root cause of all key problems.				a		
32	Key problem resolution always get public recognition.			a			
33	Leaders schedule root cause validation time.			a			
34	Leadership goes to the root cause to see and get understanding.				a		
35	Leaders participate in stress free problem evaluation.			a			

The discussion among the participants as they try to agree to a specific rating is probably as valuable as the focus their agreed to rating provides.

<u>Step 1.4 Area Self-Assessment</u>
<u>Visual Stress Free™ Evaluation</u>

The answers provide a visual spider diagram that portrays the gaps in the organizations work.

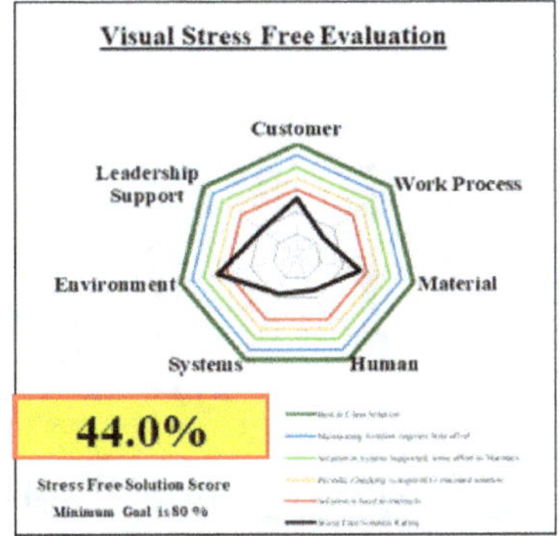

The ideal would be to have each area at the outer boundary. The heavy black line in the center represents an organization scoring itself at a 44% out of a possible 100%.

The gaps are obvious but they may not be equal in importance, or priority.

<u>Priority Rating</u>

	Prioritization				
	GAP	1-5 Business Need	1-5 Easy to do	1-5 Have Resources	Priority
Customer	4.8	5.0	4.0	4.0	384.0
Work Process	7.2	5.0	3.0	4.0	432.0
Material	4.6	3.0	3.0	3.0	124.2
Human	6.8	5.0	3.0	4.0	408.0
Systems	6.4	4.0	3.0	4.0	307.2
Environment	3.2	4.0	3.0	3.0	115.2
Leadership Support	6.2	5.0	3.0	4.0	372.0

The assessment continues with a priority rating exercise.

1. The business need for closing the assessed gap is rated on a one to five scale.
2. Next how easy will it be to close the gap and
3. Finally, are there resources that can make the improvement?

These three consideration ratings are multiplied together with the assessed gap.

<u>Opportunity Pareto</u>

The priority is graphed to produce the Opportunity Priority Pareto. In this example the number one bar is the Work Process. But the skills of the people come in a close second.

Stress Free^{TM} Work Process Improvement will solve the work process problem while at the same time greatly improving the problem solving skills of those making the improvement.

The participants of the evaluation now have much greater alignment and understanding of the improvement possibilities and the problems that may be solved. They are still at the thirty-thousand-foot level and need to get much closer to the problem.

<u>Step 1.5 Observe Selected Priority Area</u>

Material Transformation Analysis (MTA), System Diagram, Top Level Flow Diagram, Physical Layout Diagram used in step one provided a great deal of understanding. A more detailed observation of the specific problem will now take the learning in Step 1.1 and create a more detailed granularity. This much closer look with new understanding results in a dramatic increase the ability to "see".

The following tools provide this additional granularity that leads to root cause solutions.

- Travel Chart
- Time and Travel Observation Chart
- Cycle Chart
- Individual Hourly Effort Chart
- Individual minute by minute or second by second Effort Chart
- Team Effort Balance Chart

Each increases the level of problem granularity and results in a root cause solution in record time.

Travel Chart

This is used to understand the walking and the path the associate takes while doing the work. It is intended to be a pencil hand sketch of what the associated is doing.

    It:

- Will visualize the movement of people
- May include a plant layout, a department or area layout.
- It will show movement of one person or multiple people.
- It may detail the movement of products or components.
- Can be a tool to suggest layout design changes.

Example:

Before Improvement

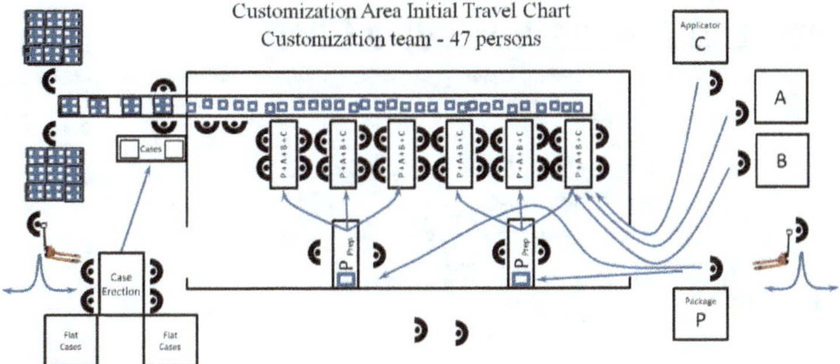

After Improvement

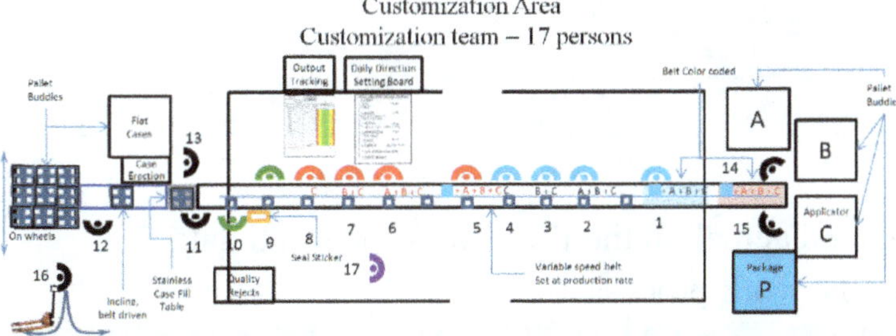

Note that only the person removing finished product and bringing materials has any travel.

In both situations the detailed motion chart for each individual is an important tool to evaluate the physical work situation of each individual.

The same conveyor was used and a Ten foot extension added.

Time and Travel Observation Chart

This is similar to the Travel Chart but intended to be more structured and less visual but with more precision.

Time and Travel Observation Process

- For the Associate being observed, record the actual time taken for each task.
- Use stopwatches. In today's world the I-phone can be used to time and record the actions of the person doing the work.
- Observe all work area team members individually.
- Repeat the observation as many times as necessary to get an accurate value.
- Determine the accurate value and record.

Example 1: Time and Travel Observation Chart

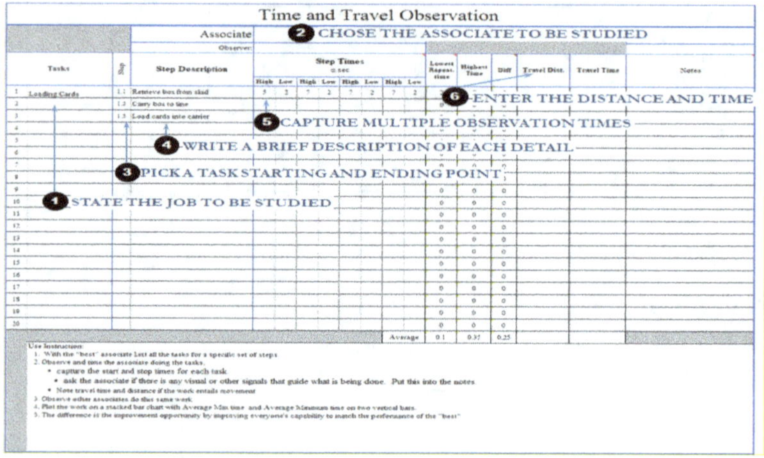

Used to:

- Understand actual on the floor work facts and data
- Clarify the work process
- See and understand the Hard and Non Value Add work – See Waste
- Understand how long each task takes to do
- Identify improvement opportunity
- Standardize work

Work Cycle Chart

The work cycle chart provides a step by step visual presentation of the length of time for each step. This provides a visual signal for potential improvement areas.

Example 1: Work Cycle Chart - Card Loading

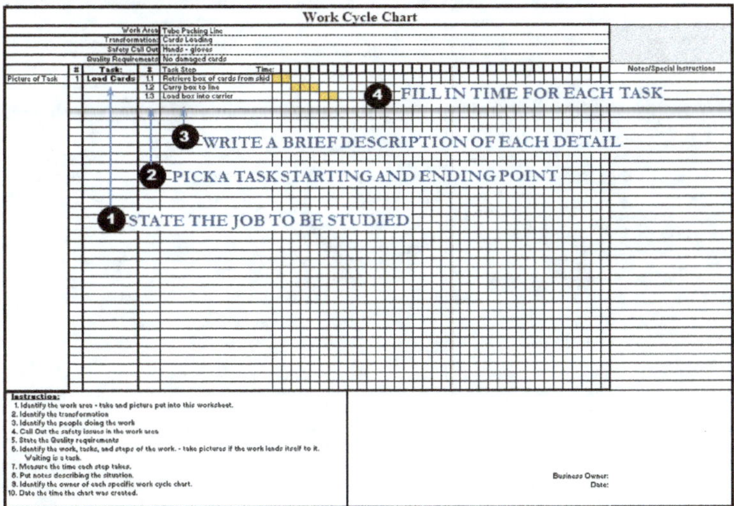

Work Cycle Time Steps

1. State the job to be studied
2. Pick a tasks' starting and ending point.
3. Write a brief description of each detail.
4. Fill in the time it takes to do each task.

Example 2: Cycle Chart

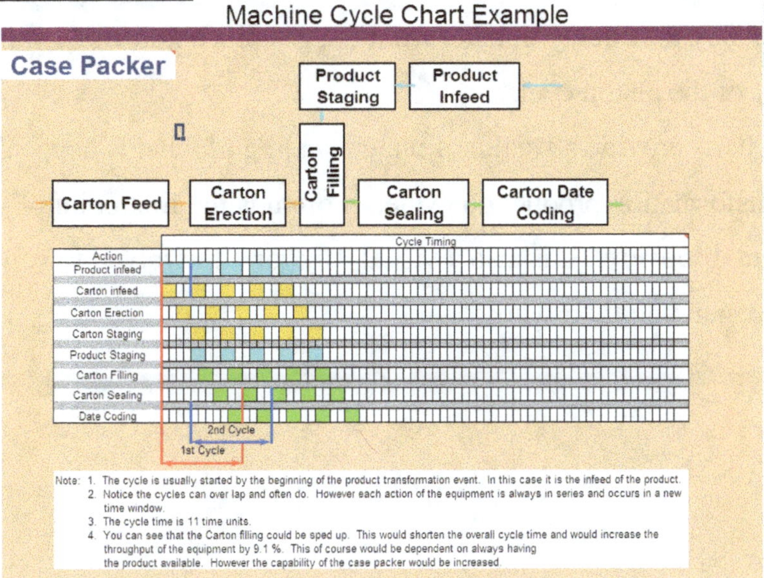

This is a time based cycle chart that relates the events in the process of packing a case of product.

A similar chart can be made studying the position of various equipment components.

The level of the study is determined by what needs to be learned or understood.

Example 3: Time based Circular Cycle Chart

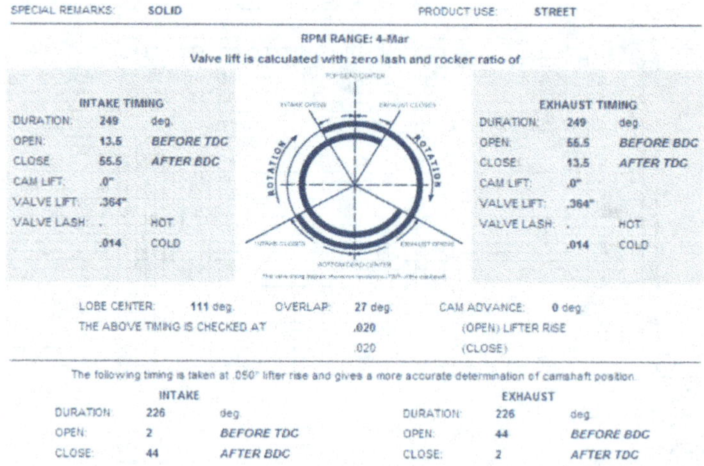

1. Note that the intake valve opens 13.5 degrees before top dead center and it closes 55.5 degrees after bottom dead center.
2. The exhaust valve opens 55.5 degrees before bottom dead center and the exhaust closes 13.5 degrees after top dead center.

This means the energy transfer of the piston is happening between 13.5 degrees after top dead center and 55.5 degrees before bottom dead center. Or said another way the energy transfer from the gasoline explosion happens in only 30.1 % of the piston cycle.

Cycle Charts provide a wealth of understanding of what is happening in the work or transformation production cycle. It often is the first point of improvement discovery. It is also a visual way for all involved to gain a deeper way of understanding their work.

Step 1.5 Observe Selected Priority Area

Individual Effort Chart - Hourly

Initially used to get a quick step by step understanding.

Example: Individual Effort Chart - Hourly

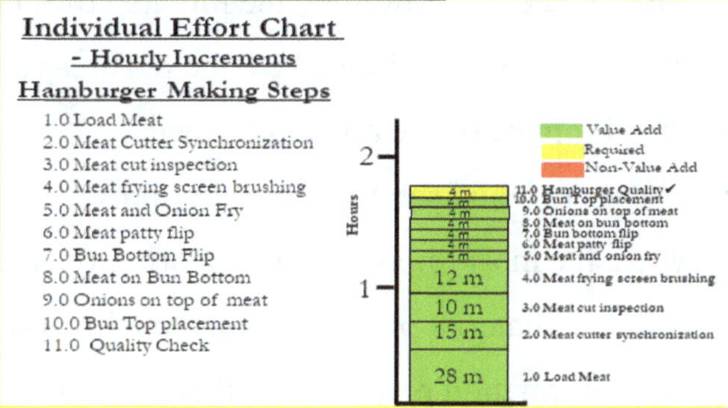

The individual effort chart captures each and every work or action done by an individual in either an eight-hour work day (480 minutes) or a twelve-hour workday, (720 minutes). This specific example uses hour by hour increments. This is only useful to get the overall material transformation. The visualization of the improvement opportunity occurs by going down to the minute versus the hourly increment.

The bar chart begins at the bottom at zero time and builds sequential to cover the entire work day. In this example the operator will execute three work cycles of the shown work sequence.

Value Add is the effort put into delivering the product desired by the customer.

One customer is the one that is paying for a product. This customer is external to the company.

However, many intermediate products get made for internal customers. In this example the next in line customer is the person that executes a quality inspection (required time), wraps and weighs the burger and moves it to the next station.

The work to satisfy either external customers or the internal customer is considered as Value Add.

Non Value is time lost to such things as waiting for materials or approval. Rework is non value add.

Required is time spent on such work as inspection to assure good quality or waiting for line clearance documentation before new product can be produced.

The build increments get finer each time an analysis is done. The one-hour effort stack is often immediately followed by the minute effort chart.

Once the entire bar for day's work is stacked, the red areas are examined to see if there is a way to eliminate or greatly reduce them.

The yellow bar is next to being scrutinized and improved.

Green is often where the most improvement opportunity is found. Material quantities, feed procedures, material placement and handling arrangements, equipment startup, and information preparation all are example improvement areas.

Individual Effort Chart Minute Scale

Each individual in the work area is tracked and all work action times are documented. A stacked bar chart provides a visual display that is easily examined for work improvement opportunities.

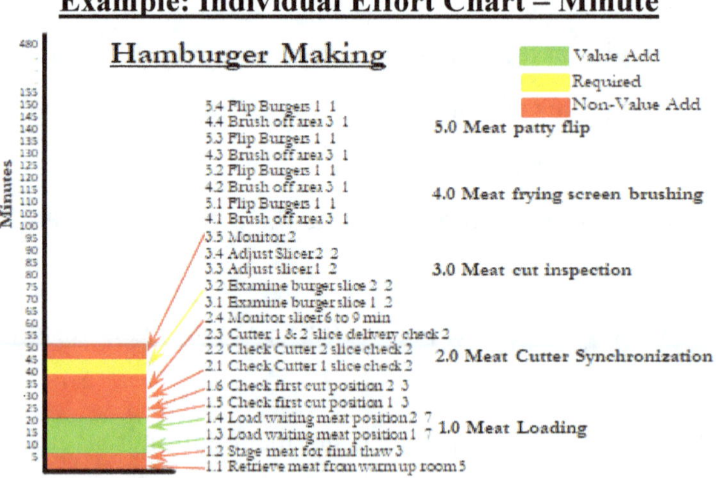

Red is the non-value add effort. Yellow is the required work that does not add value to the product such as quality inspection, or record keeping. Green is the material transformation work that adds value.

Team Effort Balance Chart

Once all individual effort charts are done they are transferred to an effort balance chart. Here the work balance among all the associates can be compared and the work adjustments made.

The variability of the work done in each of the work positions or roles is also documented. This means that multiple associates are timed at each work position. Variability must be reduced to ensure that any associate will perform equally in any position. The variability reduction comes from both training and practice.

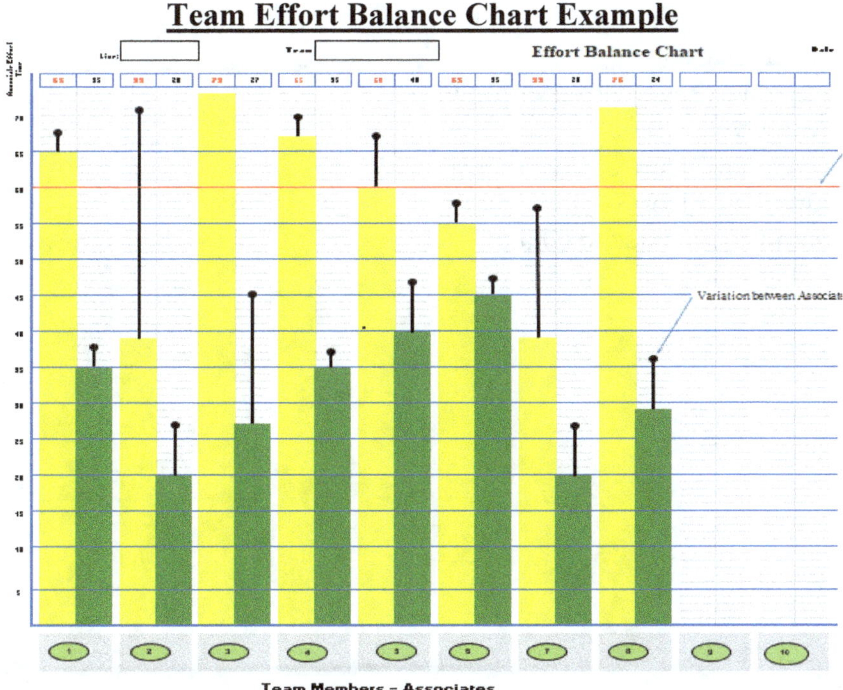

After using many or all of these ways to understand the problem the team is ready to define the improvement - Step 2: Define the Improvement.

Imagine how the initial improvement focus may have changed! It almost always is different, more focused and immediately achievable.

Step 1: Tips on Understanding the Situation

1. Spend as much time as it takes to use the suggested tools.
2. Listen to the associates doing the work. Listen for understanding.
3. Accept the different views. Have competing views work together to resolve the situation.
4. Go to the "Floor" observe the actual situation.
5. Utilize photographs and video to better see the problem.

Step 1: Tools for Understanding and Aiming

1. Material Transformation Analysis (MTA)
2. System Diagram
3. Top Level Flow Diagram
4. Physical Layout Diagram
5. Pareto Analysis
6. Travel Chart
7. Time and Travel Observation Chart
8. Work Cycle Chart
9. Individual Effort Chart - Hourly
10. Individual Effort Chart - Minute
11. Team Effort Balance Chart

Chapter 2 Document

The goal is to define the improvement in such a way that it becomes the desired way, the easy to do way for the work. The five sub steps of step 2 provide the guidance.

Step 2.1. Identify hardwork
Step 2.2. Reduce Movement
Step 2.3. Improve material flow, location, amount
Step 2.4. Mistake proof the Work.
Step 2.5. Define the VA, NVA, Required work

Step 2.1: Identify hard work

We work because we have a desire to be contributing, value adding, self-reliant members of society. The anatomy of hard work often looks similar to value added work.

People with drive overcome hard work. They however look to a plan that lessens, or eliminates the hard work. They look to eliminate the grind - the part of the work that makes in hard.

How you handle the grind is often what separates the winners from the quitters. If hard work is to be worthwhile it must meet certain standards and measures that matter to those performing the work.

Motivation sets the person up to overcome the grind of hard work. Identifying the positive elements of the motivation sets the improvement effort up for success.

Internal motivation, such as passion, self-satisfaction, personal desire is the most powerful. External Motivators such as financial reward, work security and professional recognition are also powerful. Countering these positive motivators is the fear of failure. The resentment towards those controlling the financial aspects of our lives.

Creating an improvement plan gives everyone involved an opportunity to contribute, or object to aspects of the change. Hard work is a day by day issue. The improvement needs to address the work and how it contributes to the need of the business. Why does this work need to be done? Is it value added work? Is the current struggle worth it? Identifying the improvement and making it happen quickly is important. Doing the right thing must be easier than doing the less than right thing. Everyone with an oar in the water must pull together.

The improvement plan is critical. Without it the organization will tread water. How do you identify the hard work? Use the interview technique.

Step 2.2: Reduce Movement

Physical movement, especially repetitive movement is an area to investigate. In assembly operation the same movement over and over, even when little strength is needed is an area of concern. In this case it is the number of times the same action is taken during a working period. In lifting it is the number of times, the weight and the type of movement required. In twisting to get materials the twist degree, the weight, and the frequency are factors to consider.

Motion Chart

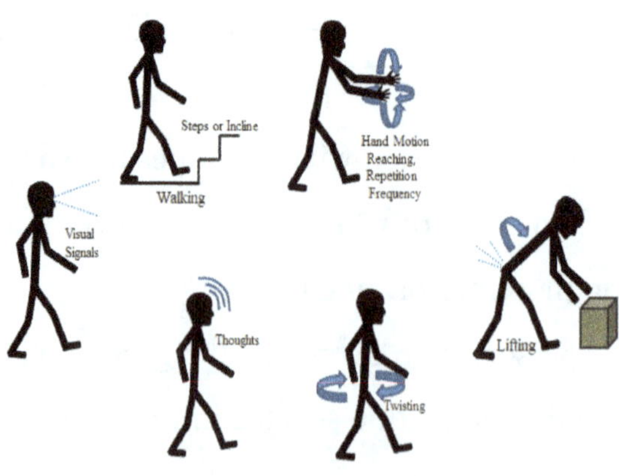

Material positioning is critical in all these cases. Assembly aids that reduce movement are another factor to consider. Work balance and rotation provides another tool to ensure hard work is reduced or shared equitably.

Walking from one side of the line to the other, going to get a tool, going to get parts or materials are all examples of movement that should be greatly reduced or eliminated.

Step 2.3: Improve material flow, location, amount

Material flow to the transformation area is crucial. The right amount, at the right time, put in the optimum position is the goal. This often requires work process change.

More frequent, smaller quantity deliveries with a just in time approach is key.

The ideal situation is for the material to exactly meet the production or work process material requirement. No material take-away at Change Over time!

Step 2.4: Mistake proof the Work

Make it impossible to do an unsafe act. Make each work process step doable in only one way. Make assembly possible only when 100% quality is assured.

Mistake proofing can be accomplished in many ways. Checklists, pictures, size gauges, dimensional guides are a few examples of mistake proofing.

<u>Examples of common Mistake Proofing</u>

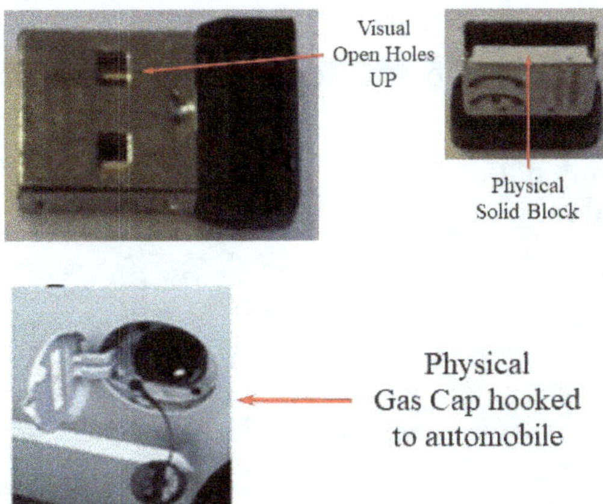

Mistake proofing is all around us. Usually it is very simple. It manifests itself in the physical world but it is also built into software and visual environments.

Step 2.5: Define the VA, NVA, Required Work

This step should constantly be on one's mind. Then as the work process is in its final stages, it is again time to evaluate the nature of the work and the various steps in the production process. Value Add, Non Value and Required work should once again be considered. Value should be optimized. Non Value eliminated or minimized. Required work should be minimized and optimized.

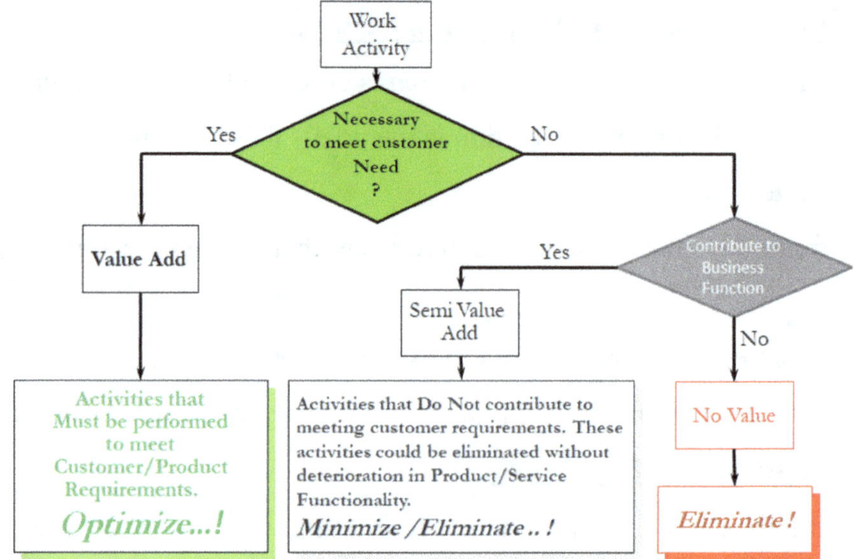

Step 2: Tips on Defining the Improvement
1. Listen to those doing the work to identify "hard work"
2. Understand that high repetition, even when it takes little strength is "hard work".
3. Detailed motion studies are critical, do them.
4. Mistake proof. It makes work easier.
5. Eliminate the red from the minute by minute work analysis.

Step 2: Tools
1. Detailed Motion Chart
2. Value Add Analysis
3. Mistake proofing

Chapter 3 Improve

Step 3 Improve

This is the take action step. Try a new approach. Think differently. Make things easy to do.

Step 3.1: Eliminate the hard work

Step 3.2: Eliminate Variation

Step 3.3: Eliminate Non-value added work.

Step 3.4: Simplify - ECRS

Step 3.1: Eliminate the hard work

In every case make the human the one that benefits. The equipment and the infrastructure should be adjusted to make it easy for the human. I have walked up to production lines where the human was expected to climb over the production line. Why not raise the production line so the human can stay on the level floor? The human is expected to work twenty, thirty, maybe forty years in this negative situation. The equipment will get modernized and renewed multiple times.

Give the human a break!

This is one example of many;

- Material adjustment points where the human is expected to climb several steps.
- The equipment is raised and the human is expected to climb several steps.
- Material feed where the human is expected to lift and place.
- Finished product where the human is expected to pick up and stack.

Utilize the Work Combination chart to evaluate the points where the human and the equipment interact. Make the production process human friendly.

<u>Work Combination Chart Example:</u>

Salami Slicer Example:

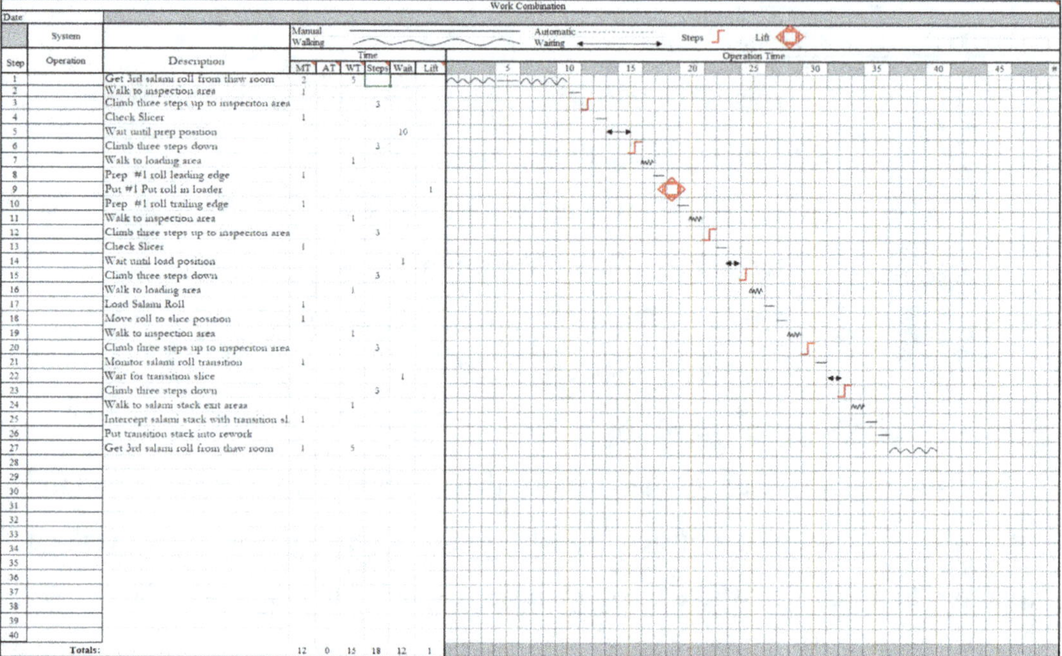

This shows the interaction of the associate responsible for sliced salami ready for packaging. The cycle time is thirty-six minutes. The time spent on each physical action is recorded. Much of this information will be available if a minute scale individual effort chart has been done.

Step 3.2: Eliminate Variation

Variation of work between different associates is a key source of work process loss. The ability to balance the work across the associates that make up the team is dependent on the ability of each associate to do the work at each work station. This means the development of every associate is important and critical to optimize and balance the work load.

Training, cross training and qualifying of all associates becomes important. Invest in the people. They will take care of the equipment. Invest in peoples' skills and they will invest their capability in improving the business results.

Training the associates in how to improve their work processes and how to solve problems to root cause are two fundamental capabilities that pay out immediately.

Training on the operation of the equipment, how work is to be done is the fundamental responsibility of the business organization.

The concept of ***mistake proofing*** is the basis of making sure that the quality of the work supports the required quality of the product. It is a partner of making work easy to do. Easy to do, mistake proofing visual instruction makes training an investment with immediate payout. Mistake Proofing makes it easy to maintain the required Product Quality

Mistake proofing prevents;
- Setup errors
- Processing errors
- Missing parts
- Wrong parts being used
- Processing omission
- Processing the wrong work piece
- Adjustment errors
- Wrong equipment setup
- Wrong tools and preparation

Step 3.3: Eliminate Non-value added work.

The time and travel chart is a tool very similar to the travel chart. At this point the elimination of non-value work is a key part of the analysis. Additionally, variability of time between associates is also of interest. The goal is minimum variability and travel distance and time.

Time and Travel Observation Example

Time and Travel Observation														
		Associate		John Henry									Date:	
		Observer:		Mary Smith										
Tasks	Step	Step Description	Observed Travel Times = sec								Lowest Repeat t. time	Highest Time	Diff	Travel Dist. Ft	Notes
			Hi	Low	Hi	Low	Hi	Low	Hi	Low					
1 Salami Slicing	1.1	Get 3rd salami roll from thaw room	720	700	700	680	720	690	710	670	670	720	50	880.0	
2	1.2	Walk to inspection area	1	0.6	1.1	0.7	1	0.8	0.9	0.6	0.55	1.1	0.55	25.0	
3	1.3	Climb three steps up to inspeciton area	1	0.6	1.1	0.7	1	0.8	0.9	0.6	0.55	1.1	0.55	3.0	
4	1.4	Check Slicer	2	1.5	2.3	1.5	1.9	1.6	2	1	0.98	2.3	1.32	0.0	
5	1.5	Wait until prep position	3	2	2	1.5	3	1.8	3	1.8	1.5	3	1.5	0.0	
6	1.6	Climb three steps down	1	0.6	1.1	0.7	1	0.8	0.9	0.6	0.55	1.1	0.55	3.0	
7	1.7	Walk to loading area	1	0.6	1.1	0.7	1	0.8	0.9	0.6	0.55	1.1	0.55	25.0	
8	1.8	Prep #1 roll leading edge	1	1	0.9	0.8	1.1	0.9	1.4	0.8	0.8	1.4	0.6	5.0	
9	1.9	Put #1 Put roll in loader	1	1	0.9	0.8	1.1	0.9	1.4	0.8	0.8	1.4	0.6	6.0	
10	1.10	Prep #1 roll trailing edge	1	1	0.9	0.8	1.1	0.9	1.4	0.8	0.8	1.4	0.6	5.0	
11	1.11	Walk to inspection area	1	1	0.9	0.8	1.1	0.9	1.4	0.8	0.8	1.4	0.6	25.0	
12	1.12	Climb three steps up to inspeciton area	1	0.6	1.1	0.7	1	0.8	0.9	0.6	0.55	1.1	0.55	3.0	
13	1.13	Check Slicer	2	1.5	2.3	1.5	1.9	1.6	2	1	0.98	2.3	1.32	2.0	
14	1.14	Wait until load position	3	2	2	1.5	3	1.8	3	1.8	1.5	3	1.5	0.0	
15	1.15	Climb three steps down	1	0.6	1.1	0.7	1	0.8	0.9	0.6	0.55	1.1	0.55	3.0	
16	1.17	Walk to loading area	1	0.6	1.1	0.7	1	0.8	0.9	0.6	0.55	1.1	0.55	25.0	
17	1.18	Load Salami Roll	1	0.7	1.2	0.6	1.3	0.7	1	0.6	0.6	1.3	0.7	8.0	
18	1.19	Move roll to slice position	1	1	1	0.6	1.2	0.7	1	0.8	0.6	1.2	0.6	3.0	
19	1.2	Walk to inspection area	1	0.6	1.1	0.7	1	0.8	0.9	0.6	0.55	1.1	0.55	25.0	
20	1.21	Climb three steps up to inspeciton area	1	0.6	1.1	0.7	1	0.8	0.9	0.6	0.55	1.1	0.55	3.0	
21	1.22	Monitor salami roll transition	2	1.5	2.3	1.5	1.9	1.6	2	1	0.98	2.3	1.32	4.0	
22	1.23	Wait for transition slice	3	2	2	1.5	3	1.8	3	1.8	1.5	3	1.5	0.0	
23	1.24	Climb three steps down	1	0.6	1.1	0.7	1	0.8	0.9	0.6	0.55	1.1	0.55	3.0	
24	1.25	Walk to salami stack exit area	1	0.7	1.3	0.8	1.1	0.8	1.5	0.7	0.7	1.5	0.8	20.0	
25	1.26	Intercept salami stack with transition slice	1	.6	1	0.8	1	0.5	1	0.6	0.5	1	0.5	3.0	
26	1.28	Put transition stack into rework	0.5	0.4	0.7	0.5	0.8	0.6	0.4	0.4	0.4	0.8	0.4	2.0	
27	1.29	Get 3rd salami roll from thaw room									0	0	0		
28											0	0	0		
29											0	0	0		
													861.0	

Use instruction:
1. With the "best" associate List all the tasks for a specific set of steps.
2. Observe and time the associate doing the tasks.
   - capture the start and stop times for each task.
   - ask the associate if there is any visual or other signals that guide what is being done. Put this into the notes.
   - Note travel time and distance if the work entails movement
3. Observe other associates do this same work.
4. Plot the work on a stacked bar chart with Average Max time and Average Minimum time on two vertical bars.
5. The difference is the improvement opportunity by improving everyone's capability to match the performance of the "best"

This cycle is 40 minutes in duration. In this example John Henry walks 17, 220 paces a day! Is that value added? Does the company really want to pay for the walking? How might this be improved in an economic way?

Step 3.4 Eliminate, Combine, Reduce, Simplify (ECRS) Evaluation

Eliminate, Combine, Reduce, Simplify (ECRS) analysis is an additional tool that is to be applied.

ECRS Mindset Questions

- **Purpose**
  - what is done?
  - why is it done?
  - what else could do it?
  - what else should be doing it?
- **People**
  - who does it?
  - why does he or she do it?
  - who else could do it?
  - who should do it?
- **System**
  - Is the computer support system 'Easy to"?
  - Is the Work System Supportive?
- **Place**
  - where is it done?
  - why is it done?
  - where else could it be done?
  - where should it be done?
- **Sequence**
  - when is it done?
  - why is it done at that time or order?
  - when could it be done?
  - when should it be done?
- **Method**
  - how is it done?
  - why is it done so?
  - how else could it be done?
  - how should it be done?

Step 3: Improvement Tips
1. Focus on making it easy for the person.
2. Utilize all the analysis observations.
3. Do the inexpensive improvements first.

Step 3: Tools
1. Work Combination Chart
2. Time and Travel Observation
3. ECRS

Ron Mueller

Chapter 4: Evaluate

Step 4: Evaluate

Ensuring the desired and targeted results were achieved is a critical step in making any improvement change. These evaluation measures should be defined before any improvement effort has been attempted. Alignment to the success measures is a critical leadership habit that needs to be cultivated.

The evaluation of successful efforts is a point in time where improvement praise is meaningful to those who have achieved the improvement.

Step 4.1: Easy To

Step 4.2 Improvement Measures

Step 4.3: Materials - Data - Information

Step 4.4: Process

Step 4.5: Systems

<u>Step 4.1: Easy To</u>

Making the work easier to do means that it will most likely be appreciated and followed. A red flag should quickly be raised when a work process is not readily followed by those doing the work. People always gravitated to the easiest way to accomplish a task. There should be no objection to doing work the easiest way possible as long as it meets quality and throughput criteria.

Is the work easier to do?

<u>Step 4.2 Improvement Measures</u>

Have the critical and predefined measures of the improvement seen a measurable improvement?

Some of the measures are;
- Safety
- Quality
- Productivity
- Cost
- Distribution
- Worker Satisfaction
- Throughput
- Overall Equipment Efficiency (OEE)
- Scrap reduction
- Customer Satisfaction

Have the critical measures improved?

<u>Step 4.3: Materials - Data - Information</u>

Getting the right quantity of materials, to the right place at the right time is of key importance. Often the timing and quantity is determined not by the process need but by the storage unit, delivery method and work balance issues. The placement of materials is often of secondary design concern.

Are materials in the right location and easy to utilize?

Are the material amounts about right?

Is Material delivery in the right time and quantities?

Is pitch synchronized? Here we define pitch as the standard Quantity of product produced with each scheduled run of the SKU.

Step 4.4: Process

Individual effort and effort balance in the work team is of critical importance. Has the work been simplified and improvements that make work easier to do implemented? Record keeping automation, visual aids, work to standards and standardized work implementation all provide process improvement.

Is duplication of effort eliminated?
Has work balance been achieved?

Step 4.5: Systems

Technology has progressed to the point that it is sophisticated and relatively inexpensive. Operational readings can now be achieved with remote reading, wireless system application. The high quantity of radio frequency interfering structures and equipment add technical challenges but walk by or proximity reading techniques are barrier busters.

Simpler system designs, slower in rate but sufficient in quantify often provide an alternate way to address multiple SKU's.

Lay out adjustments often provide ways to stream line production and improve work environment issues.

Have the systems been modified to provide better support?

Step 4: Tips

1. The earlier the success measures are identified the better. Clarity on the evaluation measures aids in the proper guidance of the improvement effort. This a key point for anyone leading the process.

Step 4: Tools

1. Measure definition
2. Field Observation
3. Control Charting of the Results

Ron Mueller

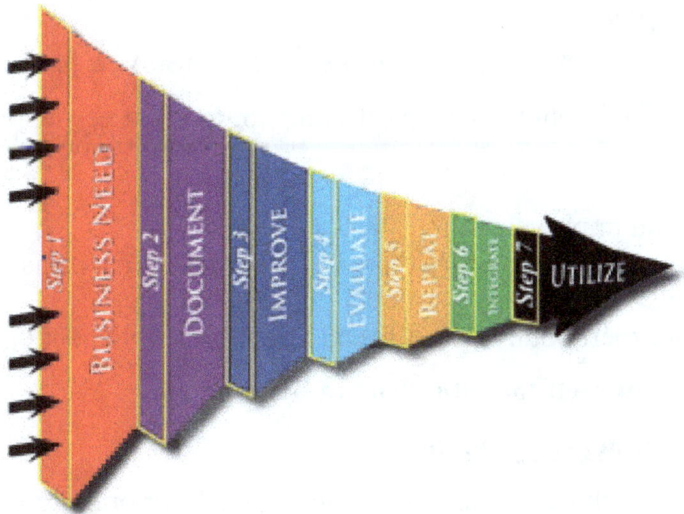

Chapter 5: Repeat

Step 5 Repeat Steps 2-4

A first cycle of improvement has been executed. Clear improvements have been achieved. The energy of the work team is high.

Don't stop.

Make the next improvement now!

The team members should have additional insights and fresh ideas. Have associates from other areas or other sights review and input their ideas.

This step often results in a break through. Something unexpected often surfaces and makes the first four steps the foundation for the breakthrough.

Step 5.1 Brainstorm

Step 5.2 Re-examine, re-evaluate.

Step 5.3 Summary Report

Step 5.1 Brainstorm

This is one of the few times that brainstorming to solve a problem is recommended. The intent is to get a different approach, a new approach to surface.

- Brainstorm improvements!!!
- Select the best ones and try them out. It is OK to try ideas and have them fail.
- Have an expert buddy watch for additional improvement opportunities.

Don't allow yourself to be the frog in hot water.

Step 5.2 Re-examine, re-evaluate.

Look over everything that has been done. Check for omission, complication, missed simplification. The work combination chart is useful for closely examining each interface.

Work Combination Chart

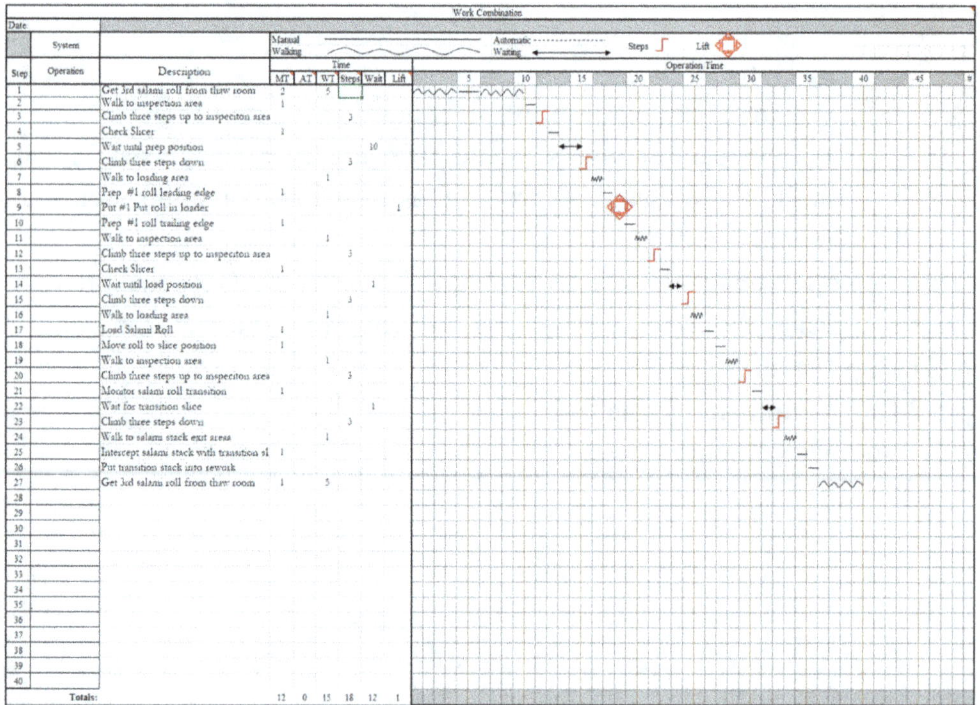

Step 5.3 Summary Report

At this step a report out to the supporting leadership is appropriate and meaningful. Use the report as the transition to step 6.

Summary Report Example:

<table>
<tr><td colspan="2" align="center">Salami at its Best</td><td>Date: Today
Associate: John Henry</td></tr>
<tr><td>Topic:</td><td align="center">Salami Slicing Improvement</td><td>Improvement $ Value: $ 75,000</td></tr>
<tr><td colspan="3">1. Purpose and Link to the business
The work in this area can be improved to be easier to do and more efficient. Productivity will be increased by twenty percent.</td></tr>
<tr><td colspan="3">2. Current situation.
The associates in this area spend: 40 % in value add and required work
32 % waiting
and   28.% walking</td></tr>
<tr><td colspan="3">3. Improvement Key Points:
- Relocated Salami thaw room. - walk time reduced by 80%

- Reduced waiting with visual controls by 57%

- Optimized Salami roll prep and loading - reduced time by 30%

+ Added Quality inspection and documentation to role

+ Added Autonomous Maintenance to role</td></tr>
<tr><td colspan="2">4. Result Data

- Walk time = 5% of work time

- Waiting time = 14 % of work time

- Value Add and Required time = 81% of work time</td><td>5. Standardize

6. Reapplication Recommendation
Evaluate other work positions and improve

Reapplication Approval
John Henry</td></tr>
</table>

Step 5: Tips

1. Try for two or three immediate cycles of improvement.
2. Strive for more granularity and detail on each cycle.
3. Take the time to document.
4. Study each previous improvement cycle's documents.

Step 5: Tools

1. All the tools described in previous chapters.
2. Summary Report

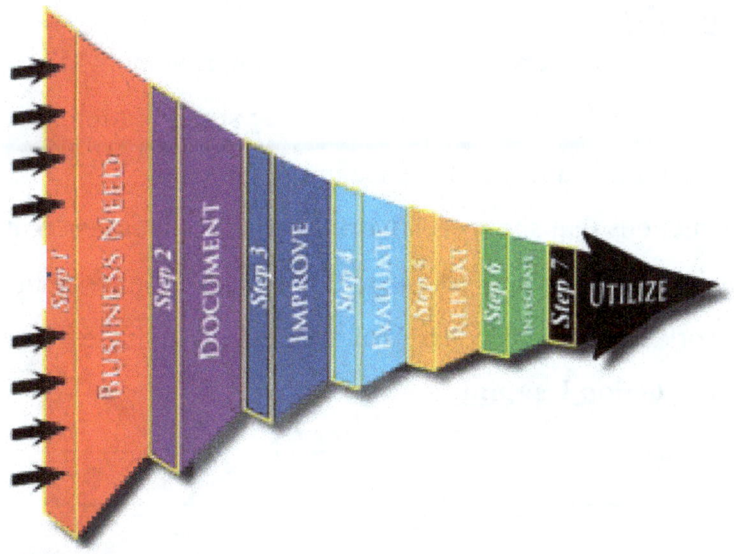

Chapter 6: Integrate

Step 6: Integrate

The improvement has been made but the decay begins immediately. Integration means making the improvement part of the work culture and practice. Maintaining the improvement must be made "easy to". Maintaining the improvement is often as challenging as making the improvement.

Step 6.1: Visualize

Step 6.2: Integrate

Step 6.1: Visualize

Visual control provides a very public way for an improvement to be maintained. Create Visualized Step by step document showing size, location, orientation of information aids in the maintenance of the improvement.

Visual instructions that show each work step makes it easier to utilize the process and quicker to train new people to do the task. The discipline to visualize the work steps creates yet a final opportunity to make improvement.

Visual Instruction Example

Visual Instruction

Process: Task:		Associate Owner:
Safety:	Quality:	

Visual Step Instruction

Step1:

Picture	Instruction	Embed Document
	Safety:	
	Quality:	
	General:	

Step2:

Picture	Instruction	Embed Document
	Safety:	
	Quality:	
	General:	

Step3:

Picture	Instruction	Embed Document
	Safety:	
	Quality:	
	General:	

Step 6.2 Integrate
Training and Qualification process

Every associate with and uses the improved work process must qualify in the flawless delivery of the product of the improved work. New members must be trained and qualified. Periodically all associates may need to requalify. All of these situations should be addressed by the training and qualification system.

Support Systems

- Safety
- Quality
- Spare Parts
- Maintenance
- HR
- Engineering
- Finance and Accounting
- Purchasing

<u>Step 6: Integrate Tips</u>

1. Enroll affected personnel in the evaluation.

<u>Step 6: Tools for Integration</u>

1. Statistical Analysis

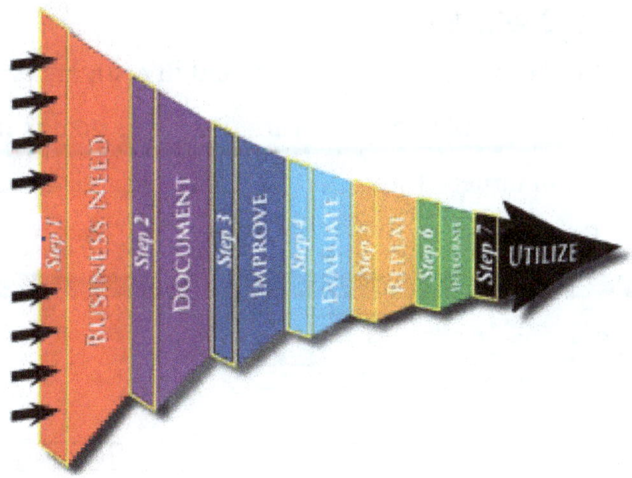

Chapter 7: Utilize
Step 7: Utilize

Step 7.1 Track Variation

Step 7.2: Daily Management System (DMS)

Step 7.3: Schedule Improvement Evaluation

Step 7.1 Track Variation

Plot the variability between team members and investigate problems/issues. Track the production output and note any variation. The goal is to stay to the hour by hour production target. Either too low or too high is unacceptable.

Output Tracking Example:

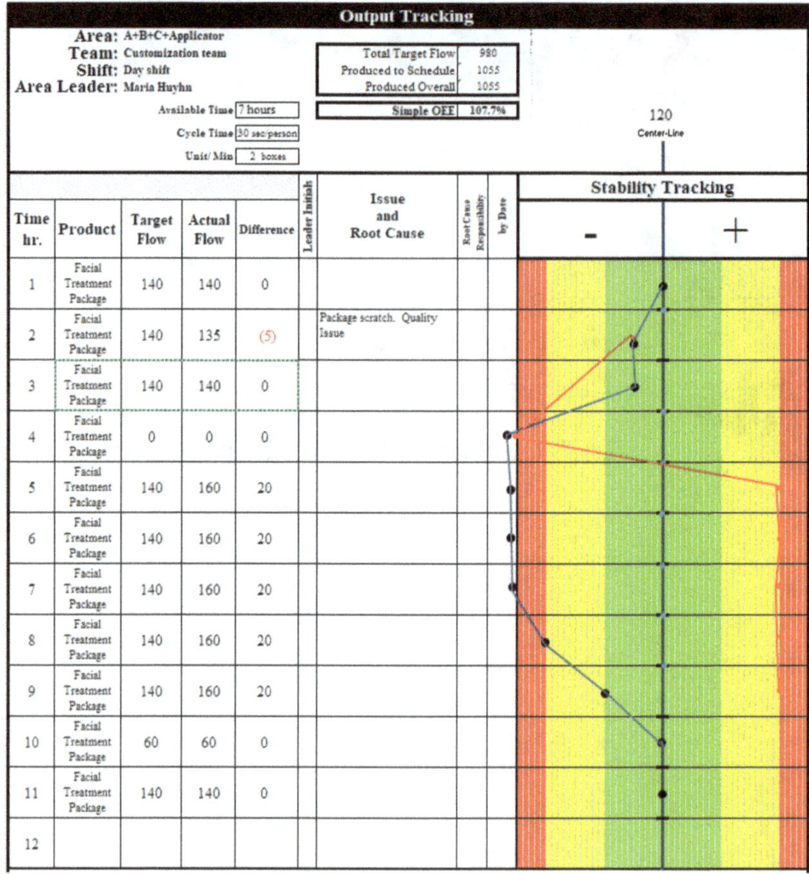

Output Tracking

Area: A+B+C+Applicator
Team: Customization team
Shift: Day shift
Area Leader: Maria Huyhn

Available Time: 7 hours
Cycle Time: 30 sec/person
Unit/ Min: 2 boxes

Total Target Flow: 980
Produced to Schedule: 1055
Produced Overall: 1055

Simple OEE: 107.7%

120
Center-Line

Time hr.	Product	Target Flow	Actual Flow	Difference	Leader Initials	Issue and Root Cause	Root Cause Responsibility	by Date	Stability Tracking −	Stability Tracking +
1	Facial Treatment Package	140	140	0						
2	Facial Treatment Package	140	135	(5)		Package scratch. Quality Issue				
3	Facial Treatment Package	140	140	0						
4	Facial Treatment Package	0	0	0						
5	Facial Treatment Package	140	160	20						
6	Facial Treatment Package	140	160	20						
7	Facial Treatment Package	140	160	20						
8	Facial Treatment Package	140	160	20						
9	Facial Treatment Package	140	160	20						
10	Facial Treatment Package	60	60	0						
11	Facial Treatment Package	140	140	0						
12										

Step 7.2: Put into the Daily Management System (DMS)

- **Practice**

  Have all the required team members practice the work process.

- **Coach and Review**

  Provide coaching for those having trouble with the process

Daily Management System Diagram Examples:

Production Area

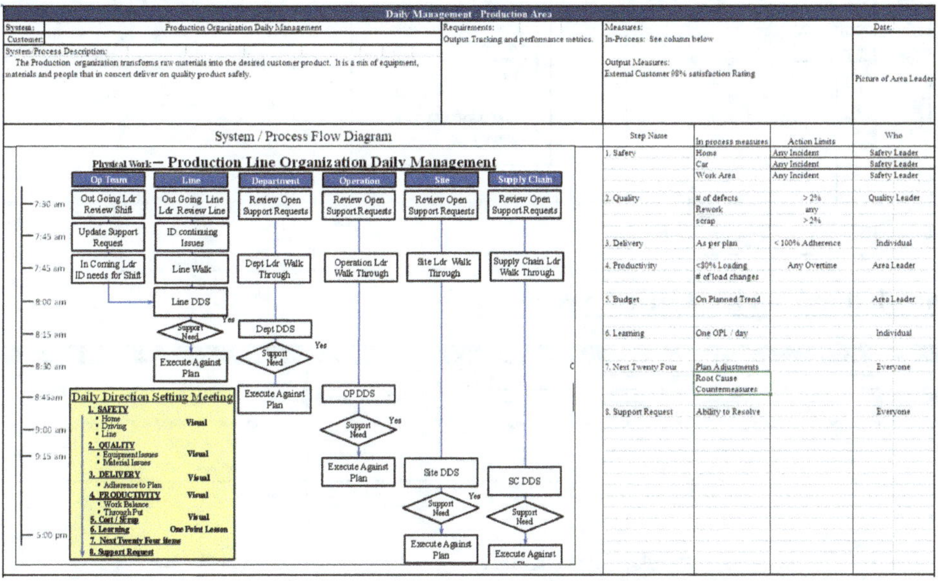

Manual Assembly Area

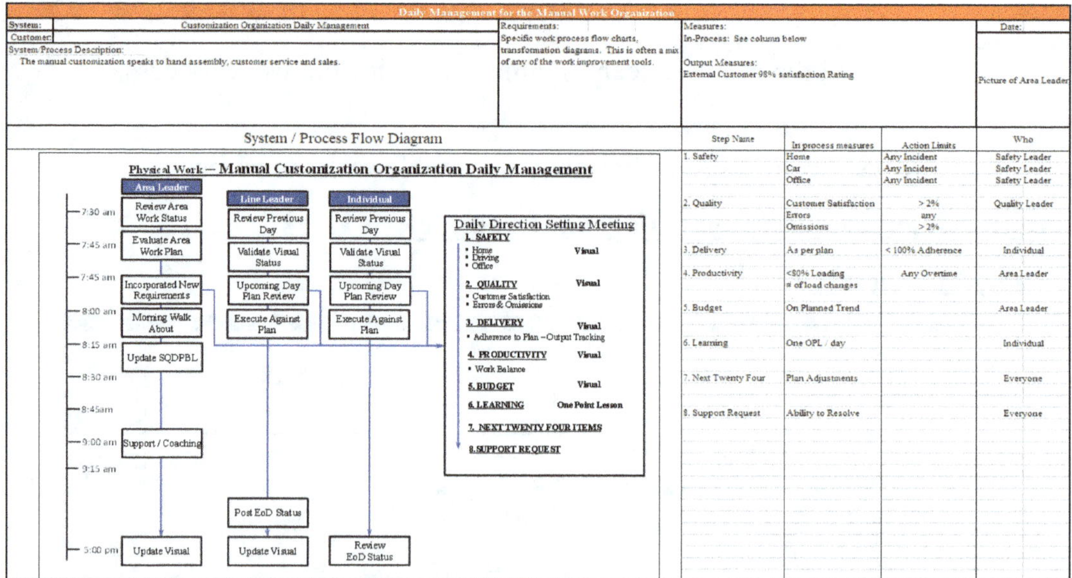

Knowledge Work – Office Area

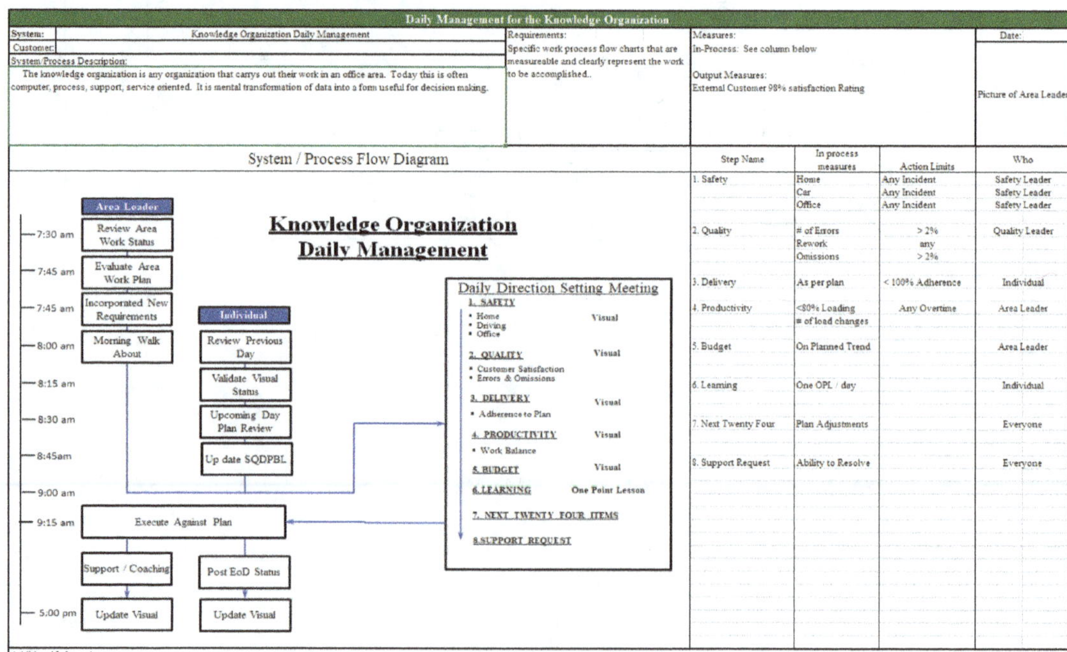

Step 7.3: Schedule Improvement Evaluation

It is critical for the organization to periodically evaluate the effectiveness and efficiency of all work processes. Evaluate what has changed. What new ideas could improve the work process?

Utilize the Stress FreeTM Work Area Self-Assessment and begin the next round of defining the improvement opportunities.

Step 7: Tips on Utilize

1. Use the hand edited field paperwork immediately. The formal paperwork may take longer to get through the system. Assign an owner to update all documents.

Step 7: Tools for Utilize

1. Maintenance Calendar
2. Retraining Schedule

Ron Mueller

Ron Mueller P.E.

- Integrated Work Systems (IWS) materials author
- Coach to dozens of Manufacturing Directors across the world.
- Certified TPM Coach.
- Tested and proven to enable true breakthrough improvement of Supply Chains.

A proven leader of smart systems implementation across supply, manufacturing and distribution to drive out cost, inefficiencies and to establish synchronized Supply Chains. He utilized the best thinking of Japan's TPM leaders and crafted the necessary related pillars and systems that work in Consumer Products Manufacturing. The results delivered include reduction of Raw and Finished Product Inventories by 40%. Delivered over $100 million is loss reduction through focused systems Workshops across dozens of sites. Developed P&G IWS program materials for external sale. Winner of P&G's Diamond Award for Contribution to Product Supply.

Core Competencies include:

- ✓ Coaching Manufacturing Leadership,

- ✓ Implementation of Integrated Work Systems,

- ✓ Statistical Replenishment design and implementation,

- ✓ Supply Chain Synchronization; author of 3 books in the Stress Free™ series that aid Business and Supply Chain leaders to develop and improve their organization's performance.

Gordon Miller P.E.
- Manufacturing Performance Program
- Development and Delivery Expert.
- Application of Intelligent Manufacturing technology against biggest business challenges with proven business results.

A record as a collaborative and leading-edge thinker, developing programs to deliver cost, productivity and growth enabling manufacturing technology systems deployed via smart standards and empowered teams. As an early developer of PR/OEE measures and improvement programs, has experience with unlocking organization capability for improvement with smart strategies. Led program that developed initial P&G Manufacturing Execution System, leveraged globally across multiple GBUs. Influenced Beauty and Household Care manufacturing systems changes that enabled and leveraged global standardization for rapid footprint growth. Experience that enabled 50% reduction in OEE losses. Experience as a leader of corporate STEM talent strategy can assess and devise approaches to ensure Talent needs for the challenging future are met.

Core Competencies include:
- ✓ Global Productivity Program Design and Management,
- ✓ Advanced Manufacturing Technology Innovation and Strategy Development,
- ✓ Development of Highly Effective Global Teams,
- ✓ Vendor development and management, Organization Capability Development,
- ✓ Talent Strategy

Ron Mueller

Stress Free™ Supply Chain Solutions

Stress Free™ Manufacturing Solutions

Stress Free™ Work Process Solutions

Stress Free™ Changeover Solutions

Stress Free™ Daily Management Solutions

AROUND THE WORLD PUBLISHING, LLC,